D1567465

BOTTOM-LINE AUTOMATION

Dedication

To Liz, Derek and Erin

Table of Contents

List of Figures

List of Abbreviations

ABC	activity-based costing
CIM	computer integrated manufacturing
CMS	cost management systems
CRT	cathode ray tube
DCS	distributed control system
DEC	Digital Equipment Corporation
DPM	dynamic performance measures
HP	Hewlett Packard
IBM	International Business Machines
I/O	input/output
JIT	just-in-time manufacturing
LCC	lifecycle cost
LCE	lifecycle economics
MIS	management information system
NPV	net present value
PDCA	plant-do-check-act
PDP	programmable data processor
PID	proportional, integral, derivative
PLC	programmable logic controller
QA	quality assurance
RFP	request for proposal
ROI	return on investment
SPC	statistical process control
SQC	statistical quality control
TQM	Total Quality Management
YEL	years of expected life

ACKNOWLEDGEMENTS

I would like to express my sincere gratitude to a number of people who significantly helped to make this book and the background research a reality.

A number of Invensys – Foxboro professionals contributed significantly to this work, including, but not limited to: Dan Antonellis, Russ Barr, Joe Barricella, Ron Beene, Barry Boyle, John Brkich, Larry Brown, Daniel Cervantes, Bob Cook, Mike Cook, Debbie Conneally, Ricardo Cortejoso, Marty Culverhouse, Garry Cusick, Trevor Cusworth, Mark Davidson, Nancy Delhommer, Donna Distante, Rik Evans, Stan Devries, John Eva, Carlos Fernandez, Joe Fillion, Kevin Fitzgerald, Jay Galasso, Dan Gilmore, Steve Gollemme, Harpreet Gulati, Bruce Henderson, Jim Hetzer, Neil Holden, Richard Howells, Fayyaz Hussain, Graham Jester, Ken Johnson, Rob Hasselbaum, Gary Hodgson, Jim Holt, Larry Hueni, Hesh Kagan, Randy Karg, Bill Ketelhut, Jeff Kissling, John Krause, John Kuenzler, Krishnakumer Kumer, Don Kylin, Larry Lablanc, Peter Lovelace, Bruce MacLeod, Leo McEvoy, Rebecca Marshall-Howarth, Bob Meadower, Janice Miller, Paul Miller, Jim Monahan, Ron Pariseau, Melanie Russell, George Sarney, Neil Schmidt, Tracey Sledge, Jack Spencer, Dick Staun, Marilyn Tarallo, Laurens van der Tang, Rick Whitmyre, Dave Williams, Lew Wright, Gene Yon and Barry Young. I am indebted to this group for their support.

A number of people from various organizations throughout the process industries also helped significantly in this effort including: Dave Adler, Wilfrido Arroyo, Rich Baker, Jeffery Baumer, Malcolm Beaverstock, George Bozant, Gene Bylinsky, Andy Chatha, Colin Christie, Dr. Robin Cooper, Bob Cox, Tony DeHerrara, Tom Fisher, Tom Fiske, Steve Furbacher, Jack Garrity, Jim Getchell, Jack Hickey, Dick Hill, Bill Huff, Mark Hughes, Wendy Johnson, David Lauer, Steve Leus, Mark Kluesmer, Lowell Koppel, Terry Landano, Owen Martin, Jose Louis Martinez, Dick McAllister, Gerard McCarry, Dan Miklovic, Jim Montague, Philippe Moro, Charles Morris, Joge Nader,Jo hn Nero, Aundra Nix, Larry Obrien, Doug Palmer, Sam Parino, Andy Peters, Ronald Root, Steve Sarnecki, Don Schuette, Ron Simmons, John Snodgrass, Greg Stedronsky, John Stolle, Len Sugarman, Donna Takeda, Tom Vollmann, Ray Walker, John Wason, Jim Wolf, and Dave Woll.

I would also like to thank those at ISA who helped with the editing and publishing of the book especially: Matt Lamoreaux, Shandra Botts, Greg Hale and Bob Burns.

I would also like to express my appreciation for the ongoing encouragement and support of my family. I thank my parents, Jack and Jean Martin have been a constant source of encouragement, my brothers and sisters and their families and Atwell and Nan Collins.

I especially thank my wife Liz and my children Derek and Erin for their ongoing love and support.

FOREWORD

A Solution Without a Problem

BACKGROUND — ECONOMIC FORCES LEAD TO DYNAMIC PERFORMANCE MEASURES

Automation systems have been used to help enable and optimize manufacturing processes for decades. When the digital computer was introduced as a viable tool for automation, process engineers often seemed to focus more attention on the tool than on the actual automation objectives. Suppliers of automation technology were typically selected on the basis of the "state-of-the-art" technological features they included in their systems rather than the improvements their automation technology could offer the manufacturing operation. Issues such as bandwidth, pixel resolution, and color palette seemed to be given greater weight than performance-enhancing software capabilities, such as automatic continuous loop tuning. Returns on automation investments were rarely, if ever, even calculated.

It was in this environment that, in the late 1980s, The Foxboro Company introduced its I/A Series system, which was designed to provide as many state-of-the-art features to the marketplace as was then possible. The Foxboro Company had invested significant amounts of money in the design and development of the I/A Series system and fully expected that it would help revolutionize process manufacturing. Unfortunately, after a number of I/A Series systems were installed and operating, an assessment showed that these systems were being implemented in the exact same way as traditional proprietary distributed control systems (DCS). These early assessments showed that most of the advanced features of the I/A Series system that had been critical during the selection process were not even being utilized.

At this point, the company commissioned an applied research program to determine how automation systems should be used to effectively improve plant performance. The premise of this exercise was that the engineers and technologists in process manufacturing operations had become too enamored of technology and were driving automation suppliers like The Foxboro Company to continually upgrade to new technological features with no consideration for what manufacturing or operational problem they would solve. Automation orders were being won or lost on features that were seldom used. Clearly, this technology-focused approach was good for neither manufacturers nor automation suppliers.

The applied research program was structured as an iterative interview process targeted at the management levels above the technologists in process manufacturing operations. In particular, plant managers, operations managers, production managers, as well as vice presidents of manufacturing, operations, and engineering were the focus of the interviews. Initially, four researchers interviewed hundreds of operations managers, plant managers, and corporate executives in over 700 different manufacturing operations in North America, Latin America, Europe and Asia over a four- to five-year period. The interviews focused on process manufacturing operations in the chemical, refining, paper, mining, food, pharmaceutical, water and wastewater, cement, clay and glass industry segments. In essence, the Foxboro program was structured to find real problems in manufacturing operations that the new I/A Series technology could address.

The iterative interview approach involved selecting a small number (approximately twenty) of managers for interviews. Once these initial interviews were conducted, the research team developed a presentation that summarized their results and conclusions. This presentation was then used to communicate the findings back to as many of the participants of the initial group as possible as well as to an expanded group. Their feedback enabled the researchers to fine-tune the results and conclusions and to gather additional information. This process was continued throughout the program with an ever-expanding group of managers.

The initial interviews did not bear much fruit. The research team was not really sure what it was searching for, and consequently the managers, although very cooperative, tended to talk in generalities. The primary information gathered in these early interviews focused on the business and market forces driving the process manufacturing industry segments. Interestingly, this first round of interviews revealed that the driving forces the managers identified did not vary much from segment to segment. All managers expressed concerns over the globalization of markets and resulting tough competitive economic environment that was just starting to emerge. "Shareholder value" seemed to be becoming more important than, or at least as important as, customer satisfaction. The picture they painted was one in which survival depended on significantly outperforming the competition economically. Though, interestingly, many of the managers predicted that corporate consolidation would accelerate because of globalization, it is doubtful any of them expected the level of consolidation through buy-outs and mergers that has occurred in the past few years.

The subsequent rounds of interviews proved to be much more valuable, and they have actually continued for over a decade now. The turning point in the interview process was the research team's acceptance of these driving forces of globalization, competition, and shareholder value. The team was then able to turn the process around and ask the managers what they were doing to address these forces within their companies to ensure that their firms would be among the survivors. The results were fascinating. Each of the managers described the programs they were investing in that they believed would help them survive, and, once again, there was an amazing consistency in the managers' answers.

The research team categorized the programs the managers were investing in into four areas. The first was technology. In spite of most of the managers' negative opinions about the efficacy of automation technology, all believed that if they fell behind in technology they would be at a competitive disadvantage. Many of them stated that they were uncomfortable about continuing to invest in such technology because of the lack of measurable benefits but that they were afraid not to.

The second investment category the managers identified involved quality. During the late 1980s, the quality movement hit North America and Europe. Japan had been utilizing statistical control and quality team approaches since the early 1950s, when W. Edwards Deming and Joseph Juran brought the quality movement to Japan. By the 1980s, it was clear that Japan had become a world manufacturing power, and many Japanese companies appeared to be outperforming traditional manufacturers both financially and in terms of quality. Western manufacturers were scrambling for ways to catch up quickly.

The third investment category was empowerment of personnel through automation technology. Several managers alluded to this, although until the past few years evidence of true employee empowerment has been scarce. Since this empowerment was enabled by the technology, this category was coupled with the first category as the program evolved.

The final investment area focused on major changes that appeared to be taking place in the financial reporting structures of manufacturing operations. These changes were initially identified under the banners of "cost management systems" (CMS) or "activity-based costing" (ABC). At first, the impact of this category on the research team was fairly low since the team came from The Foxboro Company, a technology-based automation supplier. But its importance became more and more evident as the interviews progressed.

The managers also expressed considerable concern over the disconnect they were detecting between the corporate strategies they were trying to execute and the behaviors and actions they witnessed in day-to-day manufacturing operations. Several managers were openly frustrated that they had invested millions of dollars in automation only to find that their manufacturing plants were not operating any better than they had decades earlier with much lower levels of automation. Interestingly, later interviews with the plant engineering and operations personnel of many of these plants revealed that they thought their plants were running much better as a result of automation technology and their own efforts.

As the interview process progressed it became apparent that *better* was defined very differently by management and the plant engineering and operations personnel. The interview team returned to a number of these operations merely to try to discern what the differences were. Management typically defined *better* in terms of reduced cost, increased yields, increased production and the like, while the plant operations and engineering personnel used phrases such as "ease of use," "openness," and "ease of engineering." Clearly, there was a major gap in perspective between the two groups.

Once the gap was identified, the interview process focused on how the more effective use of automation might eliminate it. During the 1980s, much work had been done in the areas of performance measurement systems, cost management systems, and activity-based costing. All of this work shared the goal of getting at the issues behind the gap between management and engineering/operations, but none of it had found a solution.

The Foxboro research team began working with Dr. Thomas Vollmann of Boston University to try to gain a better understanding of the knowledge gap issues. Dr. Vollmann had a solid understanding of these knowledge gap issues as they related to discrete manufacturing operations, and he helped identify a top-down analysis process that would start with a corporate strategy, the associated action plans, and ways of measuring the performance of the action plan and then systematically "de-compose" the performance measures throughout the operation. This *Vollmann Decomposition,* as we refer to it, enables the economic performance measures for each function, process, and activity in the organization evaluated according to the corporate strategic performance measures. If plant economic performance measures (money saved or extra money earned) are derived and determined according to this process, every function and person will be measured according to its contribution to the corporate plan. In this way, the gap in perspective between the plant-level personnel and the managers could be overcome.

The Vollmann Decomposition analysis proved to be a very powerful tool. The problem was that for performance measures to be effective in process manufacturing operations they had to be calculated in terms of the same time constant as the manufacturing process itself. If an operator is trying to manage and operate a process in real time and the performance measures are only provided on a daily basis, it will be very difficult for the measures to positively impact manufacturing performance.

Two members of the Foxboro applied research team, Dr. Malcolm Beaverstock and Peter Martin, developed and patented an approach for modeling real-time economic performance measures in process manufacturing operations. Their models are constructed from the data from the process instrumentation installed to monitor and control manufacturing processes. The resulting metrics are referred to as *dynamic performance measures* (DPMs) and can be used by operation, engineering, and maintenance personnel as decision-support tools during the real-time operation of the plant.

Over the past decade, DPMs have been installed in approximately thirty different process plants in industries such as chemical, gas, pharmaceutical, paper, mining, and food. In every case, the plant engineering and operations personnel have learned how to make decisions that result in significant ongoing gains in economic value to the manufacturing operation. DPMs, if properly derived, can thus be used as the primary metrics for continuous improvement programs in process plants. When they are, organizations work toward common objectives and enjoy significant improvements in economic impact, quality, and morale.

The remainder of this book will present the results and conclusions of the Foxboro applied research project without going through all of the iterations and changes that took place throughout the program. The material will be presented in its final form, with some supporting background information where necessary. This form reflects the inputs from all of the team's interviews as well as information from related books and professionals the team consulted. Placing the material in this larger context ensures that the reader gains a better understanding of what the managers being interviewed were trying to convey. I have chosen not to continually point out that what is being conveyed is a summary of inputs from multiple interviews. For the purposes of readability and flow, just the results are presented. Specific individuals and works consulted throughout the research process are identified either by quoting and footnoting where appropriate or by just citing the name of the source in the text.

The material presented is the state of the program at the time this book was written. It became clear to the applied research team that manufacturing performance is a very dynamic subject that will continuously evolve. As a result, the interview and data-gathering process associated with this applied research program will be ongoing.

CHAPTER 1

The Drive to Economic Performance

GLOBALIZATION — WHAT IT TAKES TO SURVIVE

Throughout the 1970s and 1980s business management forums were dominated by discussions about the increasing globalization of industry. But it was not until the 1990s that the full impact of globalization was actually realized: a very tough business environment in which only the best-prepared and best-managed companies survive. In many cases, the surviving companies have thrived to an unprecedented degree by growing both organically and through acquisitions.

In some respects, the decade of the 1990s was a buy-or-be-bought decade. The fiscally strong companies have consumed the weaker ones and often become quite large. But size alone has not been enough to prevent acquisition. Some of the larger companies in the world, even in traditionally untouchable industry segments, have been acquired in the past decade or so, sometimes even by smaller firms with fiscal strength and tough management.

Globalization has created a difficult environment marked by an incredible degree of organizational downsizing or perhaps, to be more politically correct, rightsizing. Some organizations are now doing an increased level of business with half of their original workforce. Some of the downsizing has been the result of technological advances, but most of it is motivated by a desire to reduce labor costs as much as possible without hampering productivity. The impact on the people working in these environments has been harsh. Longevity and loyalty seem to have lost their value in many organizations, and many employees feel they are regarded as mere piece-parts and numbers rather than valued contributors.

Globalization has led to several other trends that businesses must respond effectively to if they want to survive. Inconsistency and instability in the global resource base for raw materials and human resources have put new pressures on manufacturers. Those that have traditionally focussed successfully on one geographic market segment now face new competitors from different parts of the world entering their traditional market. Often for these competitors raw materials are easier and less costly to access and of higher quality, giving them favorable competitive advantage. Human resources often follow a similar pattern. Twenty years ago, developing areas of the world tended to have an abundance of human resources who were low cost but typically also lower skilled and less educated. Manufacturers in the industrialized regions had no choice but to pay higher wages for the higher skills and better education of their local workforce. But this education and skill gap is rapidly closing, presenting manufacturers in developed regions with a very challenging competitive environment. They cannot survive unless they change their traditional practices.

Another key driver created by globalization is the shift in the criteria used to select automation solutions from technical considerations to economic considerations. A decade ago almost all decisions about automation purchases depended on the preference by the technical community within the manufacturing operation of a state-of-the-art technology or feature. This resulted in a significant level of capital spending for automation technology that we now see provided little or no economic return.

Ten years ago, the phrase *shareholder value* was seldom, if ever, heard. Today, this phrase can be heard in almost every business discussion, even among the technical communities of manufacturing operations. The concern with shareholder value is introducing a new and extremely beneficial perspective into the process of selecting and using automation assets in manufacturing operations. This concern has been intensified by reductions in capital spending and capital budgets. In the 1980s, companies expected returns on capital in the range of 15 to 18 percent. Today, the focus on shareholder value has often resulted in expectations for returns on capital in the 50 percent or greater range. This drive has presented manufacturers with significant challenges in obtaining the capital they need to enhance and upgrade their technology base.

The driving economic forces that have resulted from globalization are all a great cause for concern among manufacturers. Business managers of these firms realize that to survive and thrive within this tough competitive environment they have to be better than their competitors. In the past, they could surpass competitors through superior sales and marketing strategies. In the future, they know they will have to win through more effective manufacturing strategies and better execution of those strategies.

Manufacturers have experimented with dozens of different programs, such as Total Quality Management, to set themselves apart, but most have enjoyed little success. Unfortunately, the path to survival is neither clearly defined nor without obstacles.

But the path does exist. It leads to the realization of new levels of plant performance through more effective use of all plant resources. The objective of this book is to take you down this path. Manufacturers who start along it today will likely be the world-class manufacturers of the future, the ones who survive to tell of their success.

PATHWAY TO PERFORMANCE

Manufacturers have been struggling to use technology to optimize manufacturing strategies ever since the computer was introduced as a tool for automation. From the beginning, it seemed obvious that this new digital technology offered a tremendous potential to improve on the manufacturing process. However, no matter how much effort was applied, the results never seemed to meet the expectations.

The growth in technology, especially computer technology, has resulted in some peculiar human behaviors. Technologies are typically invented to address sets of problems that humans have encountered. But the newer and more complex the technology, the more people seem to focus on the technology itself rather than the problems to be solved.

This is exactly what has happened in the manufacturing industries. Computers are ideal tools for addressing many manufacturing issues, but after they were introduced they became the object of focus rather than the problems facing manufacturers. As a result, computer technology has never reached its full problem-solving potential in manufacturing.

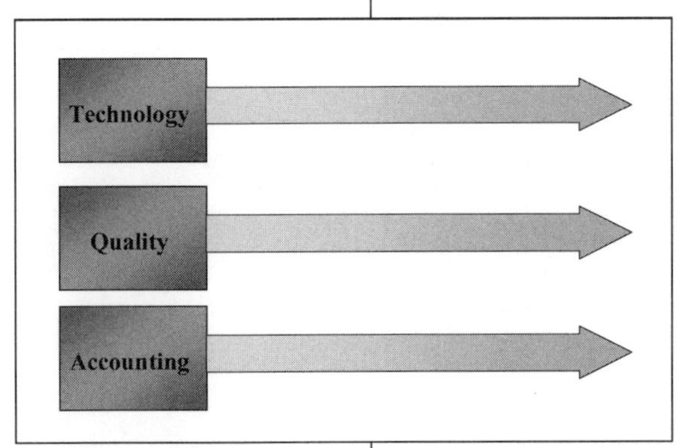

Figure 1.1 Trends for Surviving in a Global Economy

Today, much of the hype surrounding computer-based technology is behind us. Users are increasingly focused not on the tool, but on its use. This is an exciting and encouraging change, one that is vitally necessary if manufacturers are to move to a new paradigm for the management of plant or mill performance. This new paradigm is "bottom-line automation," a discipline that has the potential to significantly improve the performance of plant operations by emphasizing a complementary relationship between people and technology. If that potential is realized the result will be that ever-elusive ideal—world-class manufacturing.

One of the most startling aspects of bottom-line automation is that there is nothing terribly revolutionary about it. As Peter Drucker said in his book *Innovation and Entrepreneurship*, "the greatest praise an innovation can receive is for people to say: 'This is obvious, why didn't I think of it?'"[1] Bottom-line automation is in many respects a natural evolutionary step in manufacturing systems. It is based on three seemingly unrelated or independent trends that can actually be made to converge into a single comprehensive pathway leading to enhanced manufacturing performance.

The concept of the convergence of seemingly independent trends into a knowledge-based innovation is not new. As Drucker also pointed out, a "characteristic of knowledge-based innovations ... is that they are almost never based on one factor, but on the convergence of several different kinds of knowledge, not all of them scientific or technological."[2] In the case of bottom-line automation there are three trends (see figure 1.1). One is purely technological but the other two, "quality improvement" and "performance measurement," are not. All of three have been independently viewed by manufacturing management as part of the solution to the difficult economic driving forces we discussed earlier. Let's briefly discuss each of these trends.

THE TECHNOLOGY TREND

For about the past thirty years, industrial manufacturers have looked longingly to computer-based automation to solve their manufacturing problems. They often focused on the computer system replacing automation functions that had traditionally been accomplished with non-computer-based automation equipment such as electronic relays and/or analog controllers. In practice, however, the payback for the functional replacement of older technologies by computer-based technology never fully occurred. Doing the exact same job with a newer technology seldom provides significant returns on the investment.

Computers were introduced into the actual automation of process manufacturing plants in the 1960s. When Digital Equipment Corporation (DEC) proved that a computer could be manufactured for a reasonably low price, computer technology was seen as a viable candidate for many applications that it had traditionally not been considered for. One of the most intriguing areas of application was the manufacturing process itself.

Once installed, a primary problem faced by industry was that to be used effectively this new machine required specialists in computer technology, and these specialists were few and far between. Also, most of the people who had a reasonable knowledge of computers knew little or nothing about manufacturing. The result was that the same companies that were vendors of the traditional automation equipment performed the initial applications of computers to manufacturing plants, typically on a customized basis. Computers showed promise, but the custom application costs were very high, and just getting the computer-based systems to the same level as the previous automation systems required a great expenditure of resources.

The emphasis on computer technology in manufacturing started to evolve significantly in the 1970s. The availability of standard applications software made the computer as a tool for manufacturing automation much more viable. It allowed the computers to be *configured* to solve many of the repetitive applications found in manufacturing rather than *programmed* from scratch each time. Although the computer was becoming easier and more economical to apply, it was used, for the most part, only to replace automation functions that had previously been accomplished with older technology. Many of the manufacturing computers included software packages such as linear and nonlinear programming, but these advanced techniques were implemented sparingly. The result was that the computer was becoming an increasingly common tool for automating manufacturing facilities, but using this new technology led to few functional automation enhancements. It was simply a case of a new technology replacing an older technology.

Toward the end of the 1970s a new trend in manufacturing came into vogue, a movement to tie all of the plant's computers together. It was assumed that if all the various computers in a plant were linked together the plant would operate as an integrated facility. In manufacturing, this trend became known as "computer integrated manufacturing" (CIM). Computer integrated manufacturing became a goal unto itself in that every manufacturer, or at least every technologist in a manufacturing facility, tried to achieve some degree of "interconnectivity," and each was absolutely convinced that good things would result from interconnecting computers. Nobody seemed quite sure what these "good things" would be, but there was general agreement that, whatever they were, interconnectivity would make them happen.

Predictably, the CIM movement did not meet the high expectations of industry. In 1988, I talked to a member of a mill-wide automation team for a large pulp and paper company. He related to me that over the previous five years his company had spent over $50 million on CIM, with not one additional dollar of profit to show for the investment.

This is a prime example of the focus on the *technology* rather than the true problems in the plant. Connecting computers together into a plantwide network overcomes some of the *barriers* to implementing solutions, but it does not solve manufacturing problems. Because of its high expense, however, the CIM trend was exactly what the manufacturing world needed to drive some reality back into automation strategies and planning. Computer integrated manufacturing should have been helping plants accomplish automation, and automation should be implemented to solve plant problems. If a new development solves an automation technology problem, however, it is only of value if it results in true functional enhancements in the end application of the technology. In other words, connecting the various computers together, in and of itself, did not make the manufacturing operation perform any better.

However, it is incorrect to believe that the CIM movement had no value; it resulted in some significant advancement. Communications, operating system, and database management standards were developed during this period, and these tools helped to break down some traditional technological barriers. But the most significant positive accomplishment of this period was that it refocused manufacturers back on the functional issues of manufacturing and away from the automation technology issues.

This decade-by-decade technology drive created a tremendously fertile environment for manufacturing improvement into the 1990s. The focus on technology for its own sake is, for the most part, behind us. Technological tool sets, especially in the area of application software, are now abundant. The reduced cost of the technology now enables creative new forms of physical and functional distribution. And the emerging new automation systems have begun to take on many of the characteristics of higher-level information systems. These new tool sets provide a technology framework that is ideally suited to addressing the current opportunities facing manufacturing. Today's challenge is to take advantage of these powerful new tool sets. The time has come to apply them directly to the problems manufacturers face.

THE QUALITY TREND

The preoccupation with technology for its own sake doesn't only take place in the computer arena. It often happens when any new development emerges and seems to offer a magic cure for pressing problems. One such movement is *statistical quality control* (SQC).

Back in the early 1930s, Walter A. Shewhart, a researcher with Bell Laboratories, published a groundbreaking book on the application of statistics to the control of manufacturing processes. His book, *Economic Control of Quality of Manufactured Product* (1931), provided the foundation for a movement that would take shape some twenty to thirty years later.

Shewhart, who W. Edwards Deming referred to as "the master" of SQC, recognized that no matter how precise a manufacturing process was, in the real world it could never produce exactly the same output each time a product was made. There are variations that occur in nature that man-made manufacturing processes cannot overcome. Shewhart recognized that statistics provided an excellent platform for dealing with this natural variability. His book addressed how to apply statistical methods to the control of manufacturing processes. A good part of the fundamental concept behind the application of statistical analysis to manufactured product was that the process is "in statistical control" as long as the variability of measurement is random and within reasonable standard deviation limits. If a process is in statistical control, it is performing as well as can be expected within its natural limitations. The essence of Shewhart's theory is that *productivity improves as variability is reduced*.

W. Edwards Deming, a statistician who worked with Shewart at what is now AT&T Bell Laboratories before World War II, was very familiar with Shewart's work and is now credited with starting the quality movement in the 1950s. In 1950, Deming went to Japan and conducted a number of sessions on SQC and how this methodology could significantly improve manufacturing productivity. Because Japan was trying to recover from the impact of World War II, his message created a significant amount of interest. Obviously, many Japanese manufacturers followed Deming's advice, and his personal guidance and its results, combined with other business and manufacturing practices, have been staggering.

As Deming worked on applying statistical techniques in manufacturing plants, he developed a valuable insight: the statistics by themselves were not enough. For the statistical methods to be effective, workers had to be actively involved in the application of the SQC information. Deming became a strong advocate of worker participation and the use of structures, such as quality circles, to promote worker participation. In this way, he not only showed the value of SQC techniques; he also helped to point out that this technology applied to manufacturing processes was not enough to ensure success. There are human issues that are just as important, if not more so.

Joseph M. Juran was another of the leaders of the quality movement who had also worked with Shewart, but his focus was more on *managing* for quality improvement than on statistical techniques. Juran also went to Japan in the early 1950s, and he did much work building an entire framework for managing for quality improvement. Juran's framework is the basis for the management structures in a number of world-class corporations that have instituted successful quality improvement programs and are today realizing substantial results.

The point behind extending the quality movement from SQC to an overall management structure is not that the statistics are inappropriate, but rather that the SQC technology by itself is not enough. Today, the quality movement is vitally strong, moving ahead rapidly, and continuing to show great promise. Meanwhile, the followers of Shewart, Deming, and Juran have come to realize that the quality movement is a major management and technology movement that must be implemented as a comprehensive program to improve the economic performance of manufacturing operations. Juran's laid-back, team problem-solving approach has since been supplanted in some of the leading global companies by more rigorous performance-based approaches. One example of this is the Six Sigma program that is gaining momentum in manufacturing operations. Such performance-based quality programs have refocused many of the useful tools and concepts of statistical control into the performance-based global environment manufactures must deal with today. Though these programs are starting to show great promise, they are limited by the lack of appropriate performance-based quality indicators.

Unfortunately, some of today's manufacturing technologists continue to approach SQC techniques with a belief in technology for its own sake. As has been demonstrated many times, this approach is almost never successful. Technological advancements must have clear objectives to achieve success.

THE ACCOUNTING TREND

One of the key issues in manufacturing that has come to the forefront in the past few years has involved the way manufacturing performance has been measured. In the past, manufacturing operations often had to develop profiles of their performance to report to the outside world, generally to secure some financial backing or to report to financial backers. These systems evolved into computer-based cost accounting systems. In many respects, the first computer-based cost accounting systems were designed to report information to the financial community in the same way as earlier manual reporting systems. For a manufacturing operation, one of the traditional measures developed by this kind of system has been "cost-per-unit-product-made" (see figure 1.2). This statistic is calculated by counting the amount of each product or product type manufactured over a given period of time, calculating the cost of the operation, and assigning the costs to each product produced by the plant. Back in the early 1900s, this statistic had significance since most plants produced a very small number of different products, and a large majority of the costs in a plant were directly assignable to a product or product line. So, in a macro sense, this statistic had some historical validity and was certainly reasonable enough for the purposes of the financial community.

Since the cost-per-unit-product-made statistic was readily available from the cost accounting system and seemed to have statistical merit, it became the primary measure of manufacturing performance for almost all manufacturing facilities. Manufacturing managers were extremely cognizant of what their costs and volumes were because their pay increases, incentives, and promotions were directly driven by their operation's performance with respect to this statistic.

$$\textbf{Cost per Unit Product} = \frac{\textbf{Direct Costs + Allocated Costs}}{\textbf{Quantity of Product Produced}}$$

Figure 1.2 Traditional Product Costing

As manufacturing facilities became more complex and the number of products and product lines produced by a single facility increased, the cost-per-unit-product-made became more difficult to compute directly. Today, many production people will tell you that often less than 10 percent of their total plant costs can be directly assigned to a product or product line. This means that over 90 percent of the plant's costs have to be assigned using an allocation algorithm of some sort. This algorithm might be based on sales volume, capital expenditures, or any number of other factors associated with the products. The point is this: the statistic cost-per-unit-product-made is now rarely a valid indicator of the cost performance of manufacturing. Moreover, since manufacturing managers have little control over the numerator of that statistic, their logical goal is to make as much of their products as they possibly can in order to enlarge the denominator and thereby reduce the resulting statistic.

A new cost accounting approach referred to as "cost management systems" (CMS) has recently emerged to replace cost-per-unit-product-made. One of the primary early factors behind the development of cost management systems was to gain a better sense of the assignability of all the costs in an organization. This was accomplished through a technique known as "activity-based costing" (ABC), where all activities associated with a product or product line are accounted for and assigned to their cost profile. Functions that have been traditionally considered overhead cost functions, such as product development and sales, are now assigned to products directly. For example, if a salesperson goes to a customer's site to sell a particular product, then the cost of the salesperson's call is assigned to the product being sold.

Activity-based costing has evolved significantly since its inception. Many organizations have benefited greatly from "de-composing" traditional organizational activities that have been grouped and hidden in traditional cost accounting systems and developing cost information on an activity basis in their stead. This method has helped organizations identify activities within their operation that have little or no economic value-added. In some operations, even some manufacturing operations, these non-value-added activities could be eliminated, creating significant economic savings to the organization in areas that would probably have continued to go undetected with traditional costing approaches. However, much controversy surrounds the question of how activity-based costing may be applied in a process manufacturing environment in which the manufacturing activities are highly interrelated, connected by piping, and difficult to change. The general feeling seems to be that these techniques should apply and help add value, but few advocates have been able to determine how to accomplish this.

Early cost management systems provided a better profile of the actual cost of a facility and a reasonably accurate lifecycle cost statistic for a product or product line, but they didn't really help measure manufacturing performance. First, many of the activities associated with and assigned to a product are beyond the control of the manufacturing operation. Second, even if the statistic was perfectly accurate, the assumption would be that cost-per-unit-product-made is the correct and only statistic for measuring the performance of manufacturing, which is not necessarily the case. Third, this statistic is typically calculated after the fact and reported to manufacturing biweekly or monthly. By the time manufacturing personnel know how they are doing, either good or bad, the product that the statistics were calculated on is long gone, and there is nothing they can do about it.

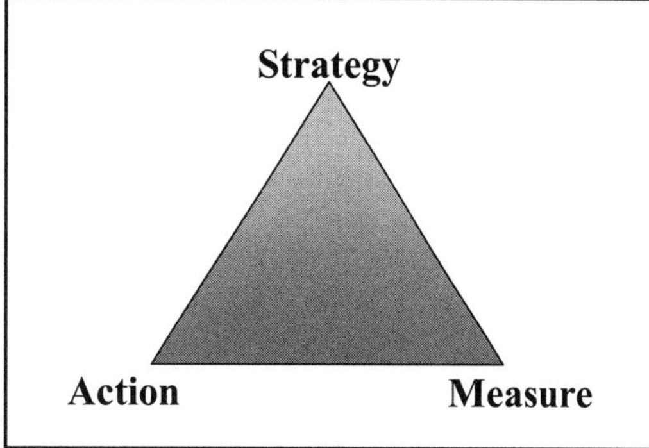

Figure 1.3 "Vollmann Triangle"

To determine whether or not cost-per-unit-product-made is the correct way to measure manufacturing performance, a simple model developed by Dr. Thomas Vollmann is quite useful. Figure 1.3 shows a triangle representing the appropriate process for implementing and evaluating manufacturing strategies. The company's manufacturing strategy is at the top of the triangle. Once the strategy has been determined, an action plan for achieving the strategy can be developed and implemented. Once the action plan is implemented, it should be measured to determine if the strategy is being met. This so-called Vollmann Triangle offers a true measure of manufacturing performance. If the strategy is strictly cost-based, then cost-per-unit-product-made may in fact be an appropriate measure. But if the strategy has any other basis, different performance measures must be developed.

Throughout the interview process for Foxboro's research program interviewees confirmed that most of today's manufacturing strategies are implemented and evaluated in two stages. First, the manufacturing strategy is developed and used to develop an action plan. Second, implementation of the action plan is initiated. But, unfortunately, the measurement of manufacturing performance is cost-per-unit-product-made, regardless of whether this statistic measures the objectives of the strategy. Just-in-time (JIT) manufacturing programs are a good recent example of manufacturing action plans that are diametrically opposed to the cost-per-unit-product-made measure. "Just-in-time" is an action plan structure based on a strategy of meeting customer demands while lowering inventory costs through the minimization of an operation's storage capacity. The goal is to make only as much of the product as the customer needs at any point in time, and no more. Note how opposed this is to the cost-per-unit-product-made statistic, which promotes making as much product in a given period as is physically possible.

The effectiveness of a JIT program should be measured in terms of customer satisfaction and the reduction of process storage. The measure must match the strategy! But, the traditional cost-per-unit-product-made measure seldom does. It is no wonder that so many new manufacturing programs fail. *Organizations tend to perform to their measures, whether or not the measures are the correct ones.*

One final point on measuring manufacturing performance is that the correct measures, once determined, must be presented to the appropriate people in the operation within a time frame that will allow for true continuous improvement. If the measures are made and presented to the operations people in charge of managing the process on an end-of-week or end-of-month basis, there will be little they can do to effect improvement. There is just too much dead time between the making of the product and the measurement of performance.

The key to success for manufacturing managers is to select the right performance indicators, institute realistic measurements, and present them to the right people on a timely basis (as the product is being made) so they can continually improve performance.

THE CONVERGENCE

The three trends we have discussed thus far—the changes in technology, quality control, and accounting—have evolved significantly in the past few years, and each of them, on its own merits, seems to offer manufacturers a ray of hope for improved performance and new paths to manufacturing excellence. In fact, let's now consider the Dynamic Performance Management (DPM) method that will demonstrate that these three trends are not unrelated but actually converge into a single comprehensive pathway to excellence: bottom-line automation (see figure 1.4).

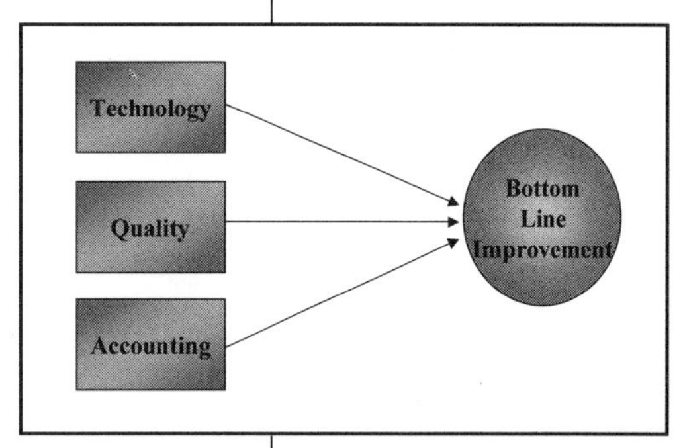

Figure 1.4 The Convergence: Survival in a Global Economy

That method, DPM, combines the essence of three seemingly independent trends into one complementary relationship. Interestingly, most manufacturers have already invested in one or more aspects of this approach, and thus the transition to a comprehensive bottom-line automation program has already begun and is now well within reach.

This convergence, by itself, is not what is important. It is the plant performance and organizational effectiveness that results from the convergence. *In today's very difficult competitive manufacturing marketplace, those manufacturers who significantly improve upon their plant performance through approaches such as bottom-line automation will become the world-class manufacturers.*

The significance of this new paradigm is that it is in fact a path toward survival! The only manufacturers who survive to see the future will be world-class manufacturers. To become world-class manufacturers, today's managers must find a way to achieve new levels of manufacturing performance. The gauntlet has been thrown at their feet.

NOTES

1. Drucker, Peter F., *Innovation and Entrepreneurship: Practices and Principles*. New York: Harper and Row, 1985, page 135.
2. Drucker, *Innovation and Entrepreneurship,* page 111.

CHAPTER 2

Technology and the Bottom Line

THE TECHNOLOGY TREND

Over the past century process instrumentation and automation technology have evolved through three major phases (see figure 2.1). Each of these phases has had a profound impact on manufacturing. To understand the current move toward bottom-line improvement, we must understand the major issues behind the transition from phase to phase. The first phase is referred to as the "Technology for Manufacturing" phase, and it coincided with the early process manufacturing operations. The second phase, "Technology for Technology," began around 1970 and continues right up to the present in the majority of process plants. The third phase is just starting to take shape and is destined to have a profound impact on the way in which automation is used. This phase is called the "Technology for the Bottom Line" and promises to lead to unprecedented economic performance improvements in the process operations that take advantage of it.

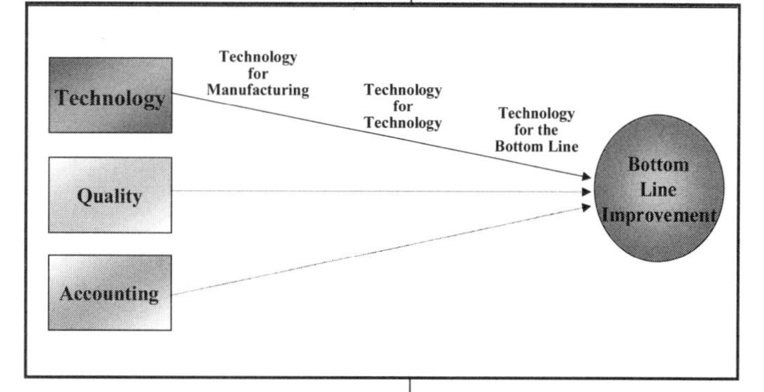

Figure 2.1 The Technology Trend

TECHNOLOGY FOR MANUFACTURING

Though digital technology is currently driving manufacturing toward a unified approach to instrumentation and automation, instrumentation and automation in the process industries and discrete manufacturing industries evolved very differently. One of the primary reasons for this divergence is that automation is not a basic requirement for discrete manufacturing. Discrete manufacturing operations involve the assembly of piece parts, which is a very visible operation that can be effectively accomplished by people. Many discrete product assembly lines began as human processes without any automation, and worked quite well. Automation was essentially introduced in discrete manufacturing to replace humans in assembly line processes with machines, such as programmable logic controllers, which were faster, never tired, and worked in a highly repeatable and reliable manner.

In process manufacturing, on the other hand, a basic level of instrumentation is required just to enable the manufacturing to take place. Many process plant parameters are totally invisible to humans because the process is hidden within piping or tanks. Instrumentation is required to make these critical parameters visible. Photographs of early process plants clearly show the gauges that were needed so plant personnel could "see" the flows, temperatures, pressures, and levels hidden within pipes, vessels, and tanks. This basic physical fact made the economic value proposition for introducing instrumentation into process plants very compelling and put the companies providing that instrumentation in a strong economic and technical position vis-à-vis plant personnel. These companies did not need to focus their energies on selling their economic value proposition because they provided manufacturing-enabling technology. Essentially, process manufacturing could not be accomplished without it.

Early process plants were primarily manually operated and employed a very basic level of monitoring instrumentation. Personnel were responsible for monitoring and when necessary adjusting valves or other actuators to manually control the processes—necessitating a large number of operators. The amount of instrumentation that a single plant employee could manage was often constrained by physical distance more than individual human capacity. Many process plants were spread out over large areas, and the instrumentation and valves could only be positioned at the point where the process was being measured or manipulated. Furthermore, many manufacturing processes involved hazardous materials. Stationing personnel close enough to use the instrumentation in hazardous environments often meant exposing them to very dangerous conditions. Plant engineers challenged the instrumentation companies to help solve these problems.

The companies responded by producing mechanical controls. The earliest controllers were typically very simple mechanical link devices. For example, level controllers similar to those still found in toilets today were sometimes used to ensure that levels in tanks did not exceed certain predefined limits. These devices did not typically improve the economic efficiency of the plant. Rather, they overcame barriers in the operation so plants could manufacture new products that had been difficult or impossible to manufacture before. The relationship between the instrumentation companies and process manufacturers was very symbiotic during this phase of the evolution of instrumentation and automation. They worked together to enable new forms of process manufacturing.

Mechanical controllers were typically limited by distance. They needed to be located and operated right at the process. This was often a difficult and dangerous proposition. Initially, the instrumentation companies addressed this challenge by introducing pneumatic controllers, which used air signals to transmit measurement information to a gauge or controller, which returned an air signal to drive the valves. These controllers could be located at a distance from the process vessels. This allowed the operators to be located in a much safer environment and oversee a much larger section of the plant by bringing multiple controllers and display stations together at a central point. Pneumatics also enabled better levels of control, making automatic control of the process more feasible. With the introduction of automatic pneumatic controllers, operators were still responsible for setting the desired operating points (set points) into the controllers, but the controllers did the second-by-second manipulation of the valves. This further freed operators to control even larger plant domains. The net result was that process manufacturers could control their plants much better with fewer people, and the plant personnel could enjoy improved workplace safety. Because of pneumatic control technology the first full-scale control rooms were implemented in process plants.

As good as pneumatic technology was for mechanical control systems, it had its own limitations. If transmission lines were very long, the nature of air signaling system meant delays in the transmission of the signals from the instrument to the controller and from the controller to the valve. These delays often led to control problems in the plants because they introduced dead time into the control loop that reduced control below desirable levels. Pneumatic systems were also expensive to maintain. Once again, process manufacturers challenged the instrumentation companies, which were now known as instrument and control companies, to help resolve these issues. In response, the instrument companies developed and introduced electronic analog control systems, which were designed to emulate pneumatic systems, but used electronic current rather than air for signaling. They were very effective and provided excellent control over much longer distances.

Whether control was effectuated manually through basic instrumentation, mechanically, pneumatically, or through electronic analog control systems, most process manufacturing managers viewed automation technology as absolutely *essential* to enabling the operation of their plant. The economic value proposition of all these systems was simple and huge. If you had them you could make product; if you didn't you couldn't make anything. This led to a very close, technology-based relationship between the instrument and control companies and process plant personnel.

During this first phase, most of the field salespeople for the instrument and control companies were professional engineers who understood the automation technology and could help customers design automation solutions using the technology. Many field salespeople spent over half their time helping customers engineer, install, and operate the automation systems. The relationship between suppliers and users was typically a very positive, consultative relationship focused on making the plant operate. The automation and instrumentation technology was an absolute necessity to make most process plants work, and thus this phase in the evolution of automation systems is referred to as the "Technology for Manufacturing" phase. Many of the managers interviewed by Foxboro's research team expressed very positive memories of the collaborative working relationship they enjoyed with the instrument and control companies during this phase. In fact, they indicated that their working relationship with these companies had changed significantly since the pneumatic and electronic analog system days, and they longed to have the old relationships back.

TECHNOLOGY FOR TECHNOLOGY

General public awareness of computer technology began to surge in the 1950s. This awareness initiated one of the most fascinating technological periods mankind has ever experienced. Computer technology seemed to become more of a cult than a science, and the manufacturing world became one of the cult's followers.

It is hard to convey the profound impact that the development of the "smart machine" had on the world to people who did not actually live through this period. Almost overnight a mythological structure developed based on the premise that computers would "take over" and run the industrial world. Manufacturing personnel gravitated into two camps: either they began to worship computers as pseudo-saviors, or they cowered in fear of the computer's awesome and mysterious capability. Meanwhile, most of the people who had originally worked in the computer field had already begun to focus much more on the computer as a technology than on the problems that technology could address.

Several key discoveries by extraordinary scientists and mathematicians led to the invention of the computer. The computer was first commercialized by Remington Rand Corporation in the early 1950s with the introduction of the UNIVAC I. IBM and Burroughs also commercialized computer systems in this period. These computers were designed for either business or scientific applications, were very expensive, and were extremely difficult to use. As a result, at first only the largest organizations in business and government represented a viable market for this new technology.

The technology behind the digital computer was a mystery to most people. The limitations of these machines were not at all clear, and many perceived the computer to be a tool of virtually unlimited power. During this period, decisions were often made to buy computers with no consideration for what the machine would actually do for the organization installing it. Many computers were acquired because "everyone else has one," and companies didn't want to be left out. With such important decisions and expenditures being made on this basis, it's clear that the focus of these organizations had obviously shifted from the business mission to the technology mission: technology for technology's sake.

In the early 1960s, Digital Equipment Corporation (DEC) overcame one of the primary barriers to the application of computers in manufacturing operations when it introduced the "minicomputer." These computers offered less computing power, but were much lower in price than the computers of the traditional vendors. Once computers became affordable, the rush was on to determine where and how they could be used. One of the obvious areas was manufacturing, and DEC's programmable data processor (PDP) product lines eventually developed an extremely large installed base in manufacturing segments.

When digital electronic technology became viable for use in process control during the late 1960s the "Technology for Manufacturing" phase came to an abrupt end. With all the advantages of the electronic analog systems, but much more flexibility, the digital process control computer was seen as the technology that would overcome all of the limitations and barriers of existing process automation.

Digital computer technology was very immature when it was introduced to manufacturing, and most people did not understand it, and to a degree feared it. These systems required new and unique skill sets that were not readily available. It is important to keep in mind that at the time the instrumentation and control technology used in manufacturing plants—relay, pneumatic, and electronic analog—was considered to be very high technology. Most instrument and control companies and most process manufacturing companies had large engineering, maintenance, and operations staffs who were familiar with the traditional technologies and were the leading technologists in their companies. But traditional control technologies required a very different set of talents than digital technology both to develop and to apply. Most instrument and control companies were forced to hire a whole new generation of employees who understood digital computer technology well, but had little experience with instrumentation or control.

These "digital gurus" carried with them a "high-tech" aura even greater than that of the existing staff, and an organizational schism often developed between the traditional instrument and control people and the new computer people. Some instrument and control companies went so far as to have two separate divisions, one for the instrument and control people—the "analog division"—and one for the computer people—the "digital division." Though the two groups clearly needed each other to apply the digital computer technology to process automation they had two very different vocabularies. The control people would talk about "PID control" and "P&IDs" while the computer people would refer to "bits," "bytes," "pixels," and "bandwidth," terms that seemed to have little to do with controlling process plants. The result of this computerization push and the schisms it caused was the most *technology-focused* period in the history of industry, one that seemed to live by the principle, "technology for its own sake."

Many plant and process engineers and operators understood the traditional pneumatic and electronic analog automation technology quite well. They knew how to design, implement, and/or operate an analog control system. Typically, however, they knew nothing at all about how to program and operate a digital computer. For computer technology to be successful, a structure had to be developed to accommodate this lack of computer familiarity. Otherwise, traditional analog control was destined to remain the furthest limit of process automation technology.

The solution was the "block" concept, in which software structures could be written to run in a real-time environment and to imitate the action of the traditional analog modules. Software control packages were now designed in terms of blocks or modules of functionality that directly matched the traditional pieces of hardware available in an analog control system. These software blocks were comprised of programs, subprograms, and files arrayed in an architecture that would support the real-time operational characteristics and functionality of process plants. Rather than the wiring that connected the analog hardware, these software blocks were "linked" together by programming code. To engineers or operators, a software "block" could essentially perform the same functions as the corresponding physical analog components, such as a PID controller, signal selector, or alarm unit, making the application of computers in process environments much more acceptable.

Now a plant's engineering staff could set up control strategies via a fill-in-the-blank configuration process, wherein each component required in the control system was activated or identified by simply filling in its specification data. Multiple blocks could be chained together to form control loops (input block + control block + output block). This new method of defining process control was pioneered by the Foxboro Company, and has been used ever since to construct continuous control strategies.[1]

To satisfy the old-school process operators, these software blocks were linked to displays that imitated the traditional analog panel boards. That is, the user interfaces the operators saw on their CRTs mimicked traditional analog control "faceplates," allowing them to more comfortably visualize their tasks.

Because the block structure allowed plant engineers and operators to use their existing knowledge when utilizing the new technology, it finally made viable the use of computers in process control. At the same time, assuming the guise of the traditional analog systems to make plant engineers comfortable with the brave new world of automation placed an artificial limit on the flexibility and potential for higher levels of automation that computers actually offered. As a result, most of the early digital control systems were implemented as exact functional replacements of either the pneumatic or electronic analog systems that preceded them. History has shown that the exact functional replacement of one technology with a newer one seldom leads to performance breakthroughs. Much of the early promise of digital computing in process manufacturing was lost.

In the 1970s, the mainstream computer industry next began to interconnect multiple computers through evolving digital communication networks. The industrial sector followed the computer industry's lead in a trend that came to be known as distributed control. The industry tended to try to avoid the word *computer* in defining these new systems because of the negative stigma computers still had for many in the industrial marketplace. This was accomplished by either giving them functional names, such as "digital process controller," or by just alluding to the fact that they were microprocessor-based components and omitting the word *computer* altogether.

Some early distributed control systems (DCS) were developed merely by networking existing process computers. As digital communication technology became accepted within industry, the next step was to move to a more functionally distributed structure in which each major control system function would have a separate computer, or set of computers, dedicated to it. A communication network interconnected the different functions. In this case, automation vendors continued to manufacture the large multipurpose process computers and make them part of the DCS structure, but they also made computer modules that were dedicated to functions such as control, process I/O, and human interfacing. The vendors also designed their own proprietary communication networks to tie all these computers together. Distributed control systems became the standard for automating the process areas of industrial plants from the mid-1970s through the early 1980s.

	Supplier A	Supplier B	Supplier C	Supplier D
Chip Set				
Bandwidth				
Protocol				
LAN Length				
LAN Media				
Pixel Resolution				
Color Palette				
MIPs				

Figure 2.2 Technology Focus

The unfortunate aspect of the evolution from early digital process computers to DCSs was that most of the innovations were on the side of computing technology. As new processors, color monitors, or communication approaches became available the automation companies would invest tremendous sums of money to incorporate them into their systems. As manufacturers tried to decide which DCS would be most suitable for their operations, the decision criteria more often than not became which supplier offered the latest state-of-the-art computer features. Many industry trade journals would run annual DCS comparison issues in which they would conduct detailed analyses to determine which DCS was the best. Figure 2.2 shows a blanked out version of an actual comparison chart right out of one of the industry trade journals of the late 1980s. It clearly shows the DCS selection criteria that journal, and many of its customers, felt were important. This figure does not attempt to show all of the criteria presented in the source article, merely a representative sample of the first high-priority criteria. Note how difficult it is to relate the criteria in the chart to whether or not the system can help operate a process plant any better. That issues such as the color palette and pixel resolution of the CRT screens were considered more important than the availability of advanced functions that might help the plant perform better is nothing short of amazing. A study of a number of requests for proposals (RFP) for automation systems made to automation companies during this same period revealed that the primary issues of concern were right in line with those presented in figure 2.2. This is a sure indicator that the people selecting the automation solutions for their plants were enamored of the computing technology and so focused on the technology itself rather than its potential to solve new manufacturing problems.

This abnormal focus on the technology of automation rather than the objectives of automation led to a long period in which significant capital investments were made with minimal, if any, returns. Process automation systems were being implemented as strict functional replacements of their analog predecessors, and the resulting plant performance improvements were very disappointing. The traditional technology-enables-solutions mindset that had been so successful with early instrument and control systems had evolved into an aberrant preoccupation with technology after the introduction of digital control automation systems. Delegating the automation decisions in process plants to the technologists, who were typically not measured in terms of plant or process economic performance improvement, helped to justify and reinforce this technology focus. This gave the technologists much power within their organizations and ensured that the selection, implementation, and operation of automation technology was going to be driven by technology objectives instead of business or financial objectives. This is not to say that the resulting decisions were always at odds with financial and business objectives, only that the technological objectives greatly outweighed them.

As more and more computer-based automation systems were installed, manufacturing management began to feel growing disappointment over the lack of tangible results. Manufacturers were investing millions of dollars in these systems, yet in many cases their plants did not run any better. For all the catchphrases and slogans bandied about to celebrate these technology-driven initiatives the promise of the "smart machine" was not being realized. A review of two of these initiatives—*CIM* and *lights out manufacturing*—may be instructive.

COMPUTER INTEGRATED MANUFACTURING (CIM) — A NOUN OR A VERB?

By the middle of the 1970s, there were many personnel in industry whose primary job was to deal with the new digital computer technology. In truth, many of them were in a unique position of influence in their organizations because not many other employees understood computers. As manufacturing management began to openly express disappointment over the new technology's lack of performance, it became clear that the jobs of these technocrats might be in serious jeopardy.

The automation technologists responded in the same way that had worked so many times before. They proposed throwing still more technology at the problem. One reason this tactic had succeeded so often in the past was that they were the only ones who understood the technology! If an operations manager spoke out against them and his lack of knowledge was made evident, his career could be severely damaged. To decide whether to adopt new technologies that they didn't understand, management turned to their technologists, even though they were the ones who stood to gain the most from these technology-focused solutions in the first place.

The new technology solution proposed in the mid 1970s was based on the fact that there were already many, many computers in factories, plants, and mills, each performing a subset of the total automation function and operating somewhat independently. If all of these computers could somehow be tied together, something "good" would surely happen. Nobody seemed to know quite what "good" was, but everybody agreed that, whatever it was, it could be made to happen. This concept became known as "computer integrated manufacturing" or, more commonly, CIM, and soon everybody was CIM-ing (to use the verb form) in one way or another.

Almost overnight, "CIM experts" appeared on the scene. They were in very high demand and, of course, highly paid. Though no two of them could quite agree on what CIM was, each of them presented equally convincing arguments supporting their points of view.

The tactic worked; very few managers were in a position to argue. Industry magazines were festooned with CIM articles. Computer integrated manufacturing was portrayed as the savior of industry and manufacturing. Everybody wanted it. Everybody could do "it." But still nobody quite knew what "it" was.

CIM IN THE PROCESS INDUSTRIES

During the CIM period little distinction was made between discrete and process manufacturing operations. The perception seemed to be that if CIM, or any major initiative, was good for one it had to be good for the other. That automation in discrete manufacturing had evolved very differently from automation in process manufacturing was conveniently ignored.

Much of the fervor associated with the CIM movement was initially directed at the discrete manufacturing industries. The automation in these segments had evolved around a work cell concept and had led to "islands of automation." The mission of CIM in these operations was to promise to interconnect the various islands and by doing so provide better logistic control over the facility.

In the process industries, which had not evolved from a work cell idea at all, CIM was nevertheless approached in the same basic way. In many process plants, however, the DCSs that had been installed encompassed much more of the plant than was encompassed by a single work cell in the discrete industries. The focus of the CIM movement in the process industries, therefore, was on connecting the DCSs to the plant's *business* systems. This was almost universally accomplished by having the DCS vendors develop gateways that would allow their DCSs to connect to the business systems in the plant (figure 2.3). Since there were many different types and vintages of business systems, even in the same organizations, an intermediate computer, often referred to as a host computer, was often employed to enable the information

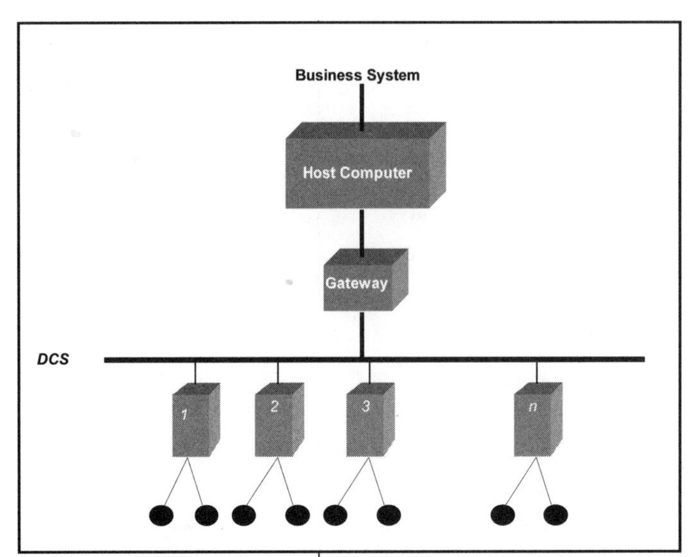

Figure 2.3 Computer Integrated Manufacturing

connectivity. DEC PDP11s, DEC VAXs, DEC MicroVAXs, HP1000s, and HP9000s probably accounted for over 85 percent of the intermediate computers in use.

Although both the business systems and the DCSs were based on digital computer technology, two very different departments were typically responsible for each in process plants. The plant engineering department was usually responsible for the DCS and the Management Information Systems (MIS) department for the business systems. In most plants, these two different departments did not work very closely together and really did not understand what the other did. In many plants visited by Foxboro's research team, before we could start a meeting involving both the plant engineering and MIS departments, we had to wait for them to introduce themselves to each other! Some general perceptions worked to create conflict and insecurity between the two groups. For example, at the time there was a general feeling that the MIS professionals should be responsible for all the computer technology in a plant. This caused the plant engineering professionals considerable concern because the MIS group typically had no notion about plant operations. The MIS group was equally concerned: they knew computers, but they had no clue about the hardware and software of DCSs. Further, MIS functions are typically transactional while process functions need to operate in "real time." Trying to get two groups that were so disjointed to work together around a common goal proved very difficult.

The net result was that the two groups often agreed to a partitioning of technical responsibilities. The boundary was typically set up as the connection between the gateway and the host computer. So it became the engineering group's responsibility to get the information across the gateway and into the host computer, and the MIS group would take it from there. Since the engineers typically had no idea what the business systems would do with the data they were providing, they tended to try to provide to the host all the data they could pull together. This was accomplished by scanning the databases within the DCS at a regular interval, typically every few minutes, and providing a snapshot of as many of the various flows, temperatures, levels, and pressures around the plant as was possible. Often a single snapshot involved thousands of points of data. This time-based record was passed through the gateway to the host computer at this time interval. The host computer would insert the record into its database. The database in the host was typically set up as a circular file with a large number of records. Each time a new snapshot was provided, the software in the host would insert it into the next available record. When the database in the host filled up, the software would insert the new snapshot record right over the oldest record in the file. In this way, at any point in time, the host's database would contain all of the values of the plant's process measurements in time slices from the present back to the number of time slices defined by the size of the database—often about a week or a month. This meant that the host had an enormous amount of data about the plant for that period.

The thinking behind this approach was that the MIS team should be able to discern anything it might need to know about the plant operations from this data. The reality was that to the MIS personnel, who really were not very familiar with plant operations, this huge data set was essentially useless information. We conducted a simple study in the late 1980s to try to determine how much of the data that had been sucked up into the host computer was ever used and found that over 95 percent of it was overwritten before it was ever accessed. If we had undertaken a more rigorous study the percentage probably would have gone up significantly. As a matter of fact, a number of plants reported that none of the data was ever accessed by the MIS team.

One other interesting anomaly about this kitchen-sink approach was the fundamental direction of the data flow. Our interviews with plant CIM teams revealed that one of the primary objectives of most CIM initiatives was to improve the plant's operational performance. In other words, the objective of the program was improving the performance of the plant in manufacturing its products. But all of the data flow was *up* to the host computer, away from the point of objective. In most of the plant sites we observed no information was being sent to the plant to make it run better. It is amazing that this fundamental design flaw was overlooked. This clearly reflects a technology-focused approach rather than a business-focused one.

One thing became abundantly clear, however, as well as universally accepted: the need to be able to connect equipment from different manufacturers together. The move was now on to develop technological solutions that could tie together different systems from different vendors. New businesses sprang up exclusively to accomplish this interconnectivity. They went by a number of different names, but were mostly referred to as "systems integrators."

Connecting disparate systems was a very expensive proposition, and automation users soon expressed their displeasure at the expense. Systems integrators and automation vendors responded by insisting that the reason for such high costs was the lack of communication standards that would make a general connectivity solution possible. The automation users accepted this argument, and in fact they responded in a much more positive way than many of the technology vendors really expected them to. A major effort was thus initiated to develop communication standards. Manufacturers and vendors alike believed communication standards would make it more cost effective to tie these computers together, and since connectivity was the essence of CIM, these connectivity standards would finally allow them to accomplish their ultimate goal at much lower cost.

The technological focus throughout this entire period was on the interconnectivity of all the computers and related equipment within a plant. The unfortunate assumption was that if all the computers were connected together, the plant would run better. Industry was obviously in for another major disappointment. The lessons of the past should have made it clear that this automation technology drive would not succeed if it did not support more automated functionality in the process of manufacturing the products themselves. Connecting equipment together does not add functionality; it merely overcomes a technology barrier to existing functionality. The evolution of these communication standards merely allowed process manufactures and suppliers to *accomplish nothing at a much lower cost.*

The entire period of automation system evolution, dating from the time the digital computer was introduced as a feasible platform for manufacturing automation in the early 1960s, has been one of technology focus rather than solution focus. The majority of CIM activities represented the culmination of this evolution, and they were, without doubt, the most extreme example of technology for its own sake ever experienced in the manufacturing sector.

As the CIM movement expanded and the fervor enveloped more and more manufacturers, experts preached the virtues of CIM through seminars, articles, and books. But to those who listened and tried to understand, it became apparent that no two experts agreed on what CIM was, and it became even less clear whether CIM was a noun or a verb.

LIGHTS OUT MANUFACTURING

The challenges posed by the global market environment have been the subject of countless conversations among manufacturing professionals. For years, most manufacturers had their market sectors largely to themselves. Not any more. New competitors from emerging industrial regions are not only entering these market segments; they are taking them by storm. They are offering similar products, often at much lower prices, and sometimes even with higher quality.

These new manufacturers are located in regions where labor costs are considerably lower than in the industrialized world. Often raw materials are readily accessible to these firms and thus much less expensive. And these new manufacturers don't even have the expense of inventing their manufacturing processes;

they can copy the same processes that the established manufacturers have already invented and developed. In some cases, they can even improve on these processes because they are starting with the very latest processing equipment.

How can traditional manufacturers compete? With high labor and raw materials costs and aging manufacturing plants, it could seem as though the battle is lost before it's even begun. Let's take a closer look at the problem; a solution may be staring us right in the face, if only we take the time to look. There is no question that if traditional manufacturers behave as they have for years, if they do not reevaluate their critical resources, the new global competitive environment will drive them out of business.

Until now, the resource that has been the brunt of the greatest criticism is the human element—in other words, labor. Ever since labor organized years ago, labor costs to manufacturers have steadily increased at a much faster rate than productivity, and in some cases productivity has actually decreased. As a result, manufacturers have been expending significant effort determining how to reduce the size and cost of the labor force. One example of this is reflected in the catchphrase of the early 1980s, "lights out manufacturing," in which technologists envisioned computer technology totally replacing thereby eliminating the requirement for any lights in the plant. The theory was that initially such technology would increase capital expenditures, but eventually it would significantly reduce the ongoing operating cost of a plant.

This approach was very consistent with the technology focus in manufacturing at the time, and just as unsuccessful. Lights out manufacturing strategies have not been realized, especially in process plants. The use of technology has led to some reductions in the labor force, but not to the expected levels, and certainly not to the extent of total replacement. No one has yet discovered how computer technology in process manufacturing can be used successfully without people. In fact, it's possible that our key to success will be to change our perspective with respect to the human element in manufacturing operations. *Those who have emphasized treating people as a resource, not a burden, have begun reaping great benefits.*

Like many of the other slogans in manufacturing of the past thirty years, "lights out manufacturing" got its start in discrete manufacturing operations. These operations began as assembly lines. As automation technology evolved, it was found that technology, such as programmable logic controllers (PLCs), could replace people on the assembly line. The PLCs were faster, cost less to operate, and were more efficient because, unlike people, they did not get tired. Replacing all the personnel on an assembly line with automation technology began to seem feasible, and the lights out movement gained momentum.

Process manufacturing plants did not evolve from assembly lines. Unlike assembly lines, where every action and response is well defined and visible, there is considerable variability in process plants—the range of things that could happen in response to a given action are manifold. For example, if the chemical composition of a catalyst or one of the reagents in a chemical reaction varies slightly from what is expected, the result of the process may be unknown. Responding to this type of variability correctly requires an intimate understanding of the operation or a level of craftsmanship that can only be provided by the experienced and knowledgeable people operating the plant. Eliminating the human element from the process operation is not as simple and straightforward as in discrete operations.

To gain a better understanding of the subtleties of applying lights out manufacturing to the process industries, we focused on the evolution of the operations personnel in process manufacturing environments. One of the best and most concise sources for this information, *In the Age of the Smart Machine* by Shoshana Zuboff, focuses on the changing level of "craftsmanship" of the labor force from the beginning of the Industrial Revolution to today. Though the scope of this excellent study is broader than just the labor force and craftsmanship, it explained the increasing quality of the labor force's output in a very clear and direct way.

In our initial interviews, the industrial managers we spoke to seldom talked about the labor force and craftsmanship or, if they did, they focused more on the high cost of the labor force than on its increased capability. Whenever we raised the issue of craftsmanship from a more positive point of view, the managers often seemed perplexed by our questions. This could have meant that they thought the fact that operations people were becoming increasingly valuable was obvious and thus did not merit discussion, or that they had never considered operations people in terms of value or quality before. Our follow-up discussions indicated that the latter was more often the case. Moreover, when the managers came to this realization, it was fascinating to see how excited many of them became.

The transformation in craftsmanship of the operators in process plants is well documented, especially in Shoshana Zuboff's book. So, we will make no attempt here to repeat it. Most readers who are familiar with the current industrial climate and conditions will easily recognize this issue.

The labor force of the industrialized world is no longer the uneducated, unskilled mass it was fifty years ago. In fact, the educational level of the populations of most industrialized countries is quite high today. If we continue to view this resource as we did half a century ago, we will certainly not be taking advantage of an indispensable resource that stands ready to help. As Thomas Peters and Robert Waterman conclude in *In Search of Excellence,* "Treating people — not money, machines or minds — as the natural resource may be the key to it all."[2]

LABOR AND EDUCATION

Before the Industrial Revolution, many of the goods on the market were produced by single-family operations. In the early stages of the revolution, factories frequently came into being when one of these family-run operations bought out others so as to consolidate resources and manufacture goods more efficiently. "The need to intensify production was the driving force behind the establishment of the early factories and workshops."[3] Typically, the family that took the initiative to consolidate became the factory's management, and the families that joined with them became the workforce. One noteworthy aspect of this arrangement was that everybody in the factory — management and labor — knew how to make the product. They were all craftsmen, and thus a high degree of craftsmanship and pride was demonstrated throughout the manufacturing operation.

This type of factory must have been a pleasure to run. The managers knew that if they or any of the workers were not available, the others in the factory could easily cover for them. Indications are that relationships between the owners and the labor force in these early mills and plants were often very close. In a way, the early success of this model was very misleading, in that it caused some people to believe that the transition to an industrial economy was going to go much easier than it actually was.

The "bathtub" curve shown in figure 2.4 provides a geometric model of the history of labor from the start of the Industrial Revolution to today. It may help to clarify the discussions to follow on the evolution of the labor movement and to identify where that movement stands today and how we can make the most of it.

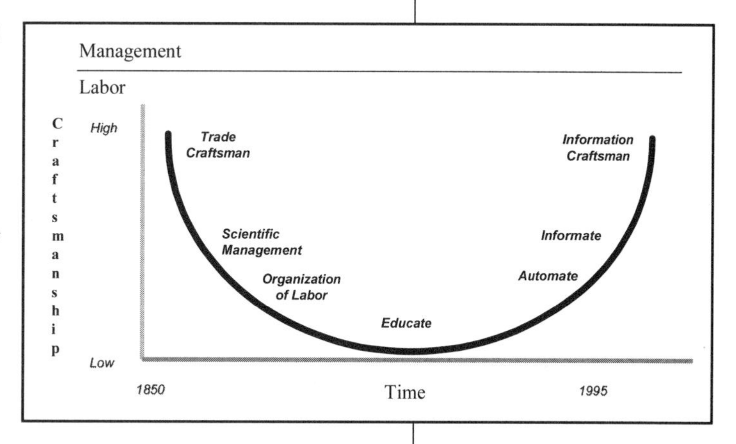

Figure 2.4 Operations Craftsmanship

As the first generation of labor began to be replaced, the individual craftsmanship of the labor force of the early family-run factories was lost. The children of these laborers had not been brought up in the factory the way their parents had and so hadn't acquired the intimate knowledge of the craft. Also, and perhaps more significantly, a major population shift was occurring from the rural agricultural areas to the industrial cities. "It took only 30 years, from 1830 to 1860, to transform both Western Europe and the Eastern United States from rural and farm-based societies into industry-dominated big-city civilizations."[4]

Almost overnight there was a huge mass of *uneducated and unskilled* people available at a very low cost. It was also clear that there were not enough highly skilled craftsmen available to effectively expand manufacturing operations. Manufacturers realized that to survive they had to take advantage of this vast unskilled, uneducated human resource.

It was a major challenge. Typically, the few skilled craftsmen who were available were recruited to join the management force in plants and mills. Thus, they would help to organize and direct the unskilled laborers. But the work habits of the new labor force were not entirely appropriate for manufacturing operations. They were accustomed to working on farms, setting their own schedules, and working somewhat independently. This mode of operation did not fit the new manufacturing environments. Workers often were late to work and took unscheduled breaks. Management tried all manner of tactics to gain control over the situation, but with limited success. "Elaborate systems of fines were developed, minutely tailored to extinguish particular expressions of the impulsive body."[5] The primary method of coercion employed by the foremen was to "combine authoritarian combativeness with physical intimidation in order to extract maximum effort from the worker."[6] But these tactics failed.

It was clear that something significant had to be done to turn this impasse into a workable manufacturing system. A more rational approach was required, one that focused on the resources available and getting the most out of them. As Zuboff notes in *The Age of the Smart Machine*, "The man who emerged as the chief symbol of the rational approach to management was Frederick Taylor."[7] The approach that he championed became known as Scientific Management, or sometimes as Taylorism.

"The agenda for Scientific Management was to increase productivity by streamlining and rationalizing factory operations."[8] In theory, this could be accomplished by applying the methods of science to the study of the manufacturing processes. This required the acquisition of new and detailed knowledge about the way in which labor worked. As Zuboff observes, "The principal method of acquiring such knowledge was the time study, later with the influence of Frank Gilbreth, the time and motion study. Here 'expert' observations of worker performance made it possible to translate actions into units of time and reconstruct them more efficiently."[9] Once this data was gathered, it, in combination "with other systematic information regarding tools and materials, laid the foundation for a new division of labor within the factory."[10]

The by-product of these studies was that management found ways "to minimize the amount of skill and training time associated with efficient operations."[11] The basic concept was to divide an entire manufacturing operation into well-defined work units and then train each laborer to do only the functions within his or her work unit. The workers did not learn how to make the entire product, or how to be craftsmen, nor did they have to as long as they could accomplish the operations in their area. Often the workers didn't even understand how their actions contributed to the overall fabrication of the product. With this limited perspective, even with years of experience the workers had little added knowledge and were no more valuable to the operation than when they had first joined.

The implementation of Scientific Management as the only reasonable approach to managing manufacturing operations marked the beginning of a strong and well-defined schism between management and labor. As Zuboff puts it, Scientific Management "legitimized a new conception of managerial responsibility in coordinating and controlling the complexities of the factory as it entered the era of mass production."[12] The laborers did the physical work with no reasonable hope of improving their lot in life. Management controlled the laborers, made much more money, and was part of a different social class. "Taylor had believed that the transcendent logic of science, together with easier work and better, more fairly determined wages, could integrate the worker into the organization, and inspire a zest for production. Instead, the forms of work organization that emerged with Scientific Management tended to amplify the divergence of interest between management and workers."[13]

The result of this situation was very predictable. The labor force had to organize to protect its interests and increase its opportunities. This organization effort took two basic forms depending on where in the world they took place: labor unions and communism. In either case, the labor force became a major political force, and as such, labor had power. One of the primary objectives of the application of this power was to make their children's opportunities greater than their own.

Many social programs evolved that focused on broadening the possibilities of the children of the labor force. One such program was public education. Although public education had been around for some time in some form, the labor movement gave it new emphasis. Workers believed that if they could get their children an education, the chances that their children would get better jobs and have better lives would significantly increase. It worked. Initially, the laborers' children became educated enough to read, write, and perform basic arithmetic. By World War II, much of the population in the industrialized world had received a secondary school education. At that time, secondary school education by itself opened up opportunities for the children of the working class to move into management positions. After World War II, it was not unusual for members of this generation to attain college degrees. This educational movement was not isolated to a few of the laborers, but represented a widespread trend. The workforce continued to become better educated, and at the same time workers became much better qualified and motivated to make significant contributions to the manufacturing environment.

As we noted earlier in this chapter, the computer was introduced into the manufacturing process in the late 1960s. Certainly in the process industries, at least, the computer has since become a primary interface between the operators and the manufacturing operations. Operators sit for hours each week "watching" the process through a "window" provided by CRTs linked to the computers. This window presents a wealth of information, on a real-time basis, concerning how the plant is running now and how it has been running over the past few hours. This technology has enabled operators to monitor and take responsibility for ever-widening areas of the plant.

Because today's operators are more highly educated, as they sit in front of their CRTs they learn about the plant in ways nobody in the operation ever has before. They watch the plant day after day, and they come to understand process relationships that nobody else understands. In essence, they are developing a *new craftsmanship*. They are becoming the information craftsmen of their age.

While working on an automation project in a process plant a few years ago, a project team was reviewing the plant diagrams to scope out the work to be done. As they reviewed the diagrams, one of the team members asked if there was any relationship between two of the process variables. The plant engineer responded that he had studied the piping and instrumentation diagrams and found that no relationship existed. One of the process operators that worked in that plant area countered that he had observed a relationship between the variables, and a small disagreement ensued. The operator contended that even though he had not studied the diagrams, he had watched the plant operate for several years and could see that the two variables reacted to each other. A more detailed study revealed that the operator was indeed correct. This is the type of nuanced, experience-based knowledge that these "laborers" bring to their jobs. It represents an entirely new form of craftsmanship.

Unfortunately, because of tradition, the information that is being presented to the majority of these craftsmen is essentially the same as was presented to their predecessors. In the meantime, the level of control exercised by automation systems has increased significantly, giving the process operators more available time. Unfortunately, the mindset of management has been to avoid giving them any more responsibility because that would mean giving up authority.

Yes, this new workforce is skilled and educated. They have built an information base on the operation of the plant that nobody else possesses. Yes, they are an expensive resource compared to labor in the developing areas of the world. But the labor force in the newly industrialized regions does not possess this new form of craftsmanship. If the manufacturers in the industrialized world want to compete with the manufacturers in the developing regions, they must use this craftsman-based labor force in an entirely different way than the unskilled and uneducated labor force is used. These operators and laborers can accomplish more than basic process and machine control; they can now be used to manage the performance of the entire plant. But while they have the skills and experience to manage plant performance, they lack two important criteria: *information and management commitment.*

Taylorism has outlived its usefulness. Today, we need a new paradigm for plant performance. We need a paradigm that makes full use of the new and valuable resources available in our plants: the information craftsmen that some still think of as mere laborers. Ideally, this new paradigm will result in the closing of the management/labor schism by empowering the labor force with information. Clearly, this would represent a very uncomfortable trend for traditional management since, as Shoshana Zuboff points out, "the prospect of shared information threatens the distinction between managers and the managed. For many managers, sharing information and maximizing opportunities for all members to become more knowledgeable is felt to be akin to treason."[14]

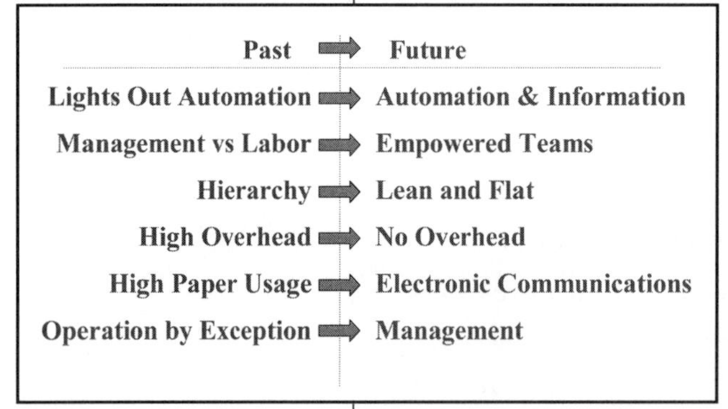

Figure 2.5 Transitions in Operations

Despite resistance, the new operations paradigm is emerging through a variety of subtle and not-so-subtle transitions now arising in many of the leading process manufacturing plants (see figure 2.5). The focus on replacing operations personnel with automation technology is giving way to a focus on providing appropriate levels of automation and information so that process operators can effectively utilize their new information craftsmanship to help drive the plant's performance. The management/labor schism is beginning to be replaced by empowered work teams or quality teams chartered with driving continuous process improvement throughout their operations. The layered hierarchical organizational structures that buried plant operations under piles of bureaucracy are being superseded by much leaner and flatter organizational structures. The overbearing overhead associated with hierarchical organizations is giving way to much more action-oriented operations. Digital computer technology is making possible the replacement of the traditional stacks of paper and reports with electronic communications that provide needed information much faster and free up operators' time to focus on performance. The net result is that the basic operator's job is changing from the traditional "operation-by-exception" model to a model of performance managers who are responsible for driving the performance of the plant.

There is one major element missing in this transition, however. If the operators are to become performance managers and drive plant performance in the right direction, they must have continual knowledge of what the plant's performance is. This type of economic performance data has traditionally been kept from operators. For this transition to the new operations paradigm to reap its true potential benefits this data must not only be made available to the operators but also in a time frame that is consistent with the job they do—and that is real time.

Given the advanced state of computer-based technologies today, real-time performance information can now be made available to everyone in an operation, right up to the front lines. Once this information is made available, once the organization's attitude is adjusted to consider the "labor force" as craftsmen and performance managers, and once authority is delegated to the workforce, plants *will* perform as never before. Concepts like "lights out manufacturing" in process plants will be relegated to the history books where they belong. And a new era of manufacturing performance will begin.

TECHNOLOGY FOR THE BOTTOM LINE

When the digital computer was first introduced to process manufacturing in the late 1960s, its promise was unbounded. Many manufacturing managers saw computer technology as the key to driving the performance of their plants to new levels and gaining real competitive advantages in their manufacturing lines. For the most part, however, after over thirty years of applying this technology in the process industries this vision has still not been realized.

As our interviews with senior manufacturing managers progressed over the past decade it became very clear that there was a major disconnect between the executives running the operations and the technologists responsible for the automation systems. When asked what their primary motivators were when they purchased automation systems, the managers stated that they were the desire to:

- Improve plant quality
- Improve safety
- Increase manufacturing flexibility
- Improve operations reliability
- Improve decision-making
- Improve regulatory compliance
- Increase product yields
- Increase productivity
- Increase production
- Reduce manufacturing costs

Although few would argue with this list, these criteria are seldom taken into consideration during the actual purchase of an automation system or during the system's life cycle.

Most of the listed criteria have a direct impact on the ongoing economic performance of the manufacturing operation. In spite of this, in most cases the only economic factor associated with the use of computers in process plants has been *price and price alone*. Even at first glance, however, it is clear that price is but one of the economic variables involved in the assessment of the economic returns of an automation system. When the decision-making and major capital investments for ongoing improvement processes are left to technologists, price typically becomes the only economic variable in the equation. This price focus has turned industrial automation systems into commodities in the eyes of the manufacturing operations. As commodities, the value of the technology is only assessed when obsolescence makes the ongoing cost of keeping the system running unacceptable. This commodity focus is a major contributing factor to the disappointing economic returns that have generally resulted from automation.

In recent years, there has been a growing trend among manufacturers to try to develop ongoing relationships with automation suppliers. Although volume price discounts have been one key aspect of these relationships, they have also often caused the manufacturers and automation suppliers to take into account a larger range of economic factors than just price. The first new factor to be scrutinized was lifecycle costs. That is, instead of just analyzing the current price of a system, manufacturers would estimate the entire cost of the system over its useful life span (see figure 2.6). The basic equation for analyzing lifecycle cost is as follows:

$$\text{LCC} = \text{Price} + \text{Project Engineering} + \text{Installation} + \text{NPV (Ongoing Annual Costs)}$$

where

LCC = Lifecycle Costs

Price = Automation system price

Project Engineering = Total cost to engineer the project

Installation = Total cost to install the system (including start-up)

NPV = The net present value function of the ongoing costs

Ongoing Annual Costs = The annual engineering, operations, and maintenance cost of the system

The net present value function calculates the time value of money over extended periods.

This lifecycle cost perspective challenged automation suppliers to provide features in their automation systems that would reduce the initial engineering and installation costs as well as the ongoing operation and maintenance costs. Generally, lifecycle costs tend to be high at the beginning of the life cycle due to the expenses of initial engineering, installation, and the product's purchase price. After start-up, they level off for several years, but then start to rise toward the end of the lifecycle because of the costs associated with obsolescence: repairs, retraining, and other

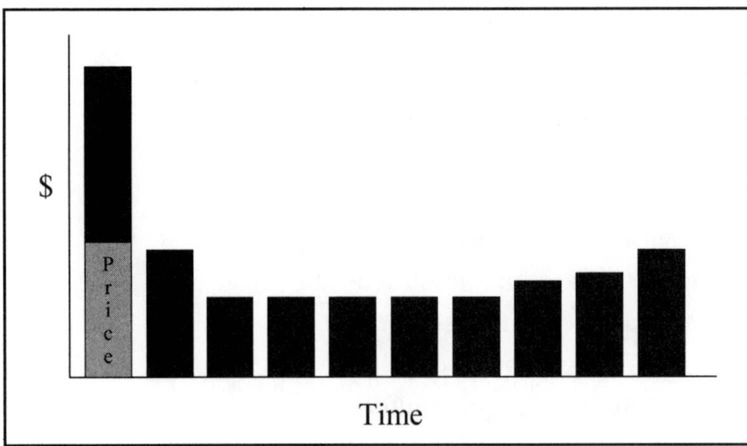

Figure 2.6 Lifecycle Cost Profile

maturity costs. Note that the system price tends to be a reasonably small component of the overall automation system cost. In fact, studies have placed the average product price at less than 35 percent of a typical project's cost, without even taking into consideration the ongoing costs. This demonstrates one of the deficiencies of the price-only approach.

The expansion of economic perspective from price to lifecycle costs was a major step forward for industry, but that perspective was still very limiting. A cost-only focus tends to relegate the automation system to the category of a "necessary evil." From this perspective, the only economic justification for upgrading or replacing a system is that the cost of maintaining the current system is too high because it has become obsolete. This is the perspective of most industrial automation users today; most of them still use a *cost-only* economic evaluation standard. If there is no perceived economic benefit to the manufacturing operation from installing an automation system then that system is certainly not meeting management's objectives, and probably nothing but the most rudimentary system should ever be deployed.

Many automation system users have talked about determining the returns on their automation investments, but to calculate returns you need more than the lifecycle costs: you need to have an accurate measure of the lifecycle *benefits* the manufacturing operation derives from using the automation system. In the interviews we conducted with manufacturing executives, we found none who were actually measuring the benefits created by automation. Most admitted that they did not know how to get at these metrics in any reasonable way and that their current cost accounting systems did not provide detailed enough economic data to be able to infer the benefit value.

The economic benefits automation systems create for manufacturing operations essentially occur in two major areas. The first is the manufacturing cost savings that result from reduced power consumption, reduced raw material costs, and reduced manpower requirements. The second is the increase in production that can be gained through better asset utilization. Of these, the only one that manufacturers regularly monitored was the reduced manpower automation made possible, because it is relatively easy to measure. The other elements of the benefit calculation are variables that constantly change as the products are being produced and are, therefore, very difficult to measure.

If all these economic benefit variables were measured, the return on the automation investment could be determined through the following equation:

Economic Return on Automation = Lifecycle Benefits – Lifecycle Costs

This equation represents the classic investment profile for any capital investment offering a return (see figure 2.7). In theory, the benefit of most capital investments should be capable of being managed so that they provide either constant returns over the life cycle or, as in the case of automation systems, continuously increasing economic benefits over the life cycle.

As the concept of lifecycle economic profiles for automation systems began to gain recognition during the late 1990s, the author was asked to lead a session on the subject at the 1996 ISA (The Instrumentation, Systems, and Automation Society) Technical Conference. At that conference and at subsequent meetings, a number of professionals from companies such as E. I. DuPont, General Foods, Eli Lilly, and Dow Chemical contributed much data from several automation projects that helped us build a lifecycle economic profile. A high-level view of

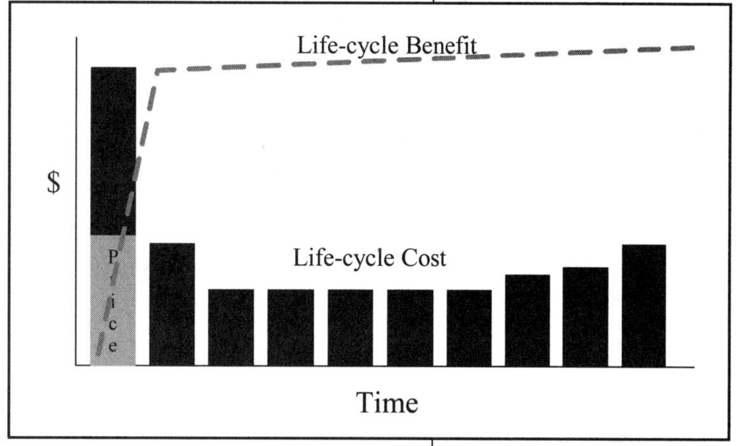

Figure 2.7 Lifecycle Economic Profile

the profile we used is shown in figure 2.8. This profile defines the benefit of an automation system as the net present value (NPV) of the annual manufacturing cost savings and the annual production increases that result from the automation system. The net present value is a function that calculates the current value of money paid on a regular basis over a number of years. It is appropriate when trying to make an up-front decision about the best economic value offered by a number of possible choices that will either be generating economic costs or benefits for a set period. An automation system does both. YEL (years of expected life) is an indicator that the NPV calculation should be done over the number of years that the automation system is expected to operate in the plant. This helps to determine the value of automation systems that may have different expected life cycles once installed. The lifecycle cost calculation has two basic components: the project costs and the ongoing costs. The project costs can be captured by the three general categories of price, initial engineering costs, and installation costs. The ongoing costs can also be calculated using the net present value function; they include annual engineering costs, annual operations costs, and annual maintenance costs.

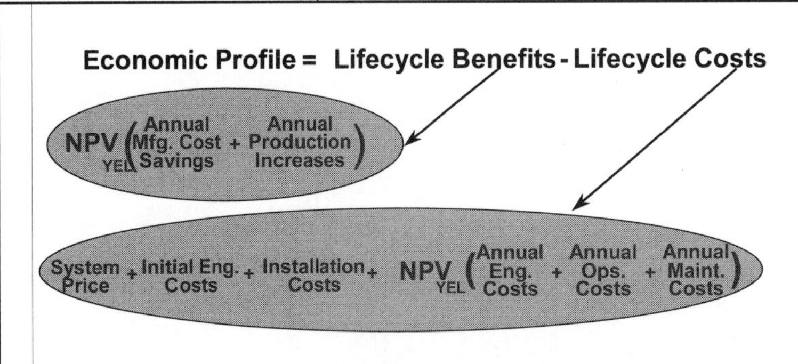

Figure 2.8 Lifecycle Economic Model

As part of my data-gathering project I analyzed several automation projects and from the data collected an actual economic profile emerged. Considerable data was available for the cost side of the profile, but only limited data was available to analyze the benefit side. This is not surprising given the lack of focus industry-wide on the benefits component of the equation. The analysis of this data produced a number of interesting results. First, as figure 2.9 depicts, most of the projects for which the benefit side was measured at all, showed a continuous decline in the benefit value over the system's life cycle. This result was shared with several of the professionals who contributed the data; they were not overly surprised by it. Many of them attribute the continuous benefit decline to the gradual obsolescence of the system. This conclusion was later demonstrated to be false since many projects do realize continuous improvement in the economic benefits derived from the automation system. In fact, the "decline" seems to be associated more with the lack of measurements of the benefit side than an actual decline caused by obsolescence. One of the basic principles of process control is that if you can't measure the controlled variable, you can't control it. The same is true for financial variables such as lifecycle benefits. It also appears that if the variable is not measured, it will most probably move in the wrong direction, which is exactly what the data showed.

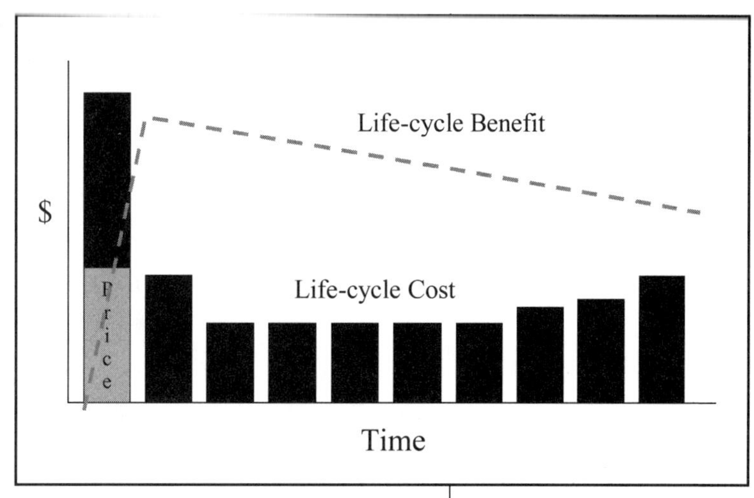

Figure 2.9 Lifecycle Economic Profile Results

Another interesting aspect of the data we collected was that the lifecycle cost data was actually distributed across the set of data, as shown in figure 2.10. For example, the price, which had traditionally been the primary economic variable in the automation system decision process, represented less than a quarter of the cost of the first five years of system use. Between the initial engineering costs and the five-year lifecycle engineering costs, the engineering of the automation system accounted for 37.8 percent of the five-year costs—considerably greater than the purchase price of the system. Perhaps the most interesting result of the analysis came on the benefit side of the model. The benefit data collected, which was not statistically valid due to the small sample size of projects for which benefit data was available, showed that the benefit-to-cost ratio over the first five years was 3.4:1. This means that the average return on automation technology for projects for which the benefit data was measured was very strong. The analysis team did not believe that the 3.4:1 ratio was repre-

	System Price	Initial Eng. Costs	Installation Costs	Annual Eng. Costs	Annual Ops. Costs	Annual Maint. Costs
Average first 5 year % of cost	23.2%	28.5%	16.1%	9.3%	7.6%	15.3%

Figure 2.10 Lifecycle Cost Breakdown

sentative of the average return on automation realized in industry. In fact, the automation users who were measuring the benefit that automation provided were considered to be among the best performers. This means that the 3.4:1 ratio is a "best-practices" result rather than an average result.

When this data was shared with a larger number of automation users, the consensus seemed to be that these results were not surprising. In fact, most of those interviewed readily accepted the results. This caused us to ask why more users weren't focused on the economic benefit created by automation. Most of those we asked did not have any specific response, but when we followed up with a more detailed analysis we uncovered three industry practices that are significant barriers to an effective total lifecycle economic approach to automation.

The first barrier was the *replacement automation* approach to automation projects, which is almost universally used in industry. This approach begins at most industrial plants when their capital budget for an automation system upgrade is approved. At that point, a project team is typically established to determine the specifications for the new automation system to be installed. In most of these cases, the specification is developed by looking at what the currently installed system does and then building the new system specification around it. This leads to requests for proposal (RFP) seeking competitive bids from automation suppliers, requests whose specifications exactly describe the old system already in place. The suppliers know that the one who will win the order is the one who meets the specification at the lowest price. Thus, even though they may have added many performance-enhancing capabilities to their computer-based automation systems since the old system was installed, they propose the lowest cost system they can. Typically, this means that all of the advanced capability the suppliers have invested in is left out of the proposal. The net result is that the new system being installed is an exact functional replacement for the system being retired. Replacing old technology with new technology that does the exact same thing seldom leads to breakthrough improvements.

When this phenomenon is pointed out to the user's project team their response is typically, "don't worry, once the new system is up and running we will take full advantage of all of the advanced capabilities." This is when the second barrier to a total lifecycle economic approach to automation starts to take effect: the *project team approach.* Project teams of highly qualified engineers are established when capital budgets for automation projects are approved. The project team works on the project throughout the project cycle. Once the project is initiated, the project team goes away. Some of the members may go on to other projects. Some may remain in the plant to manage ongoing engineering activities. But the human resources required to take advantage of the advanced features of the automation system are now no longer available. As a result, the "later on" that everyone was hoping to take advantage of never happens. The initial benefit, if any, provided by an automation system is typically the best benefit that will be realized from the system over its life cycle. From that point there is a continuous degradation of benefit.

The good news in all of this is that while this continuous degradation of economic performance goes on, nobody knows it. This is because in most cases *the benefit side of the economic profile is not measured.* This is the third barrier to success. Measurement systems are typically only valued if they are providing good news. The news that might be provided by a measurement system that continuously communicates the economic benefit due to automation would probably not be very positive. But since these measurements are not being made, most manufacturers do not feel the pain of poor performance. In this case, no news is good news. Or is it? This approach was fine when process manufacturers could not help but make significant profits no matter how good – or bad – their operations were. But in today's tough global business environment, this approach is just not acceptable.

One promising result of all of this analysis was that when the professionals in industrial operations paid attention to the economic benefit measures—even if poorly calculated on an infrequent basis—they were able to realize phenomenal results. What if these econometrics could be made and provided to the operations personnel in real time? How much better could the operation perform? Research has shown that the 3.4:1 benefit-to-cost-ratio is a very conservative indicator of what could actually be accomplished through better decision-making based on better and more timely information. The real-time econometrics of plant performance are called Dynamic Performance Measures (DPM). We will address the implementation and use of DPMs later in this book.

NOTES

1. Shaw, William T., *Computer Control of Batch Processes*. Cockeysville, MD: EMC Controls, 1982, page 100.
2. Peters, Thomas, and Robert H. Waterman Jr., *In Search of Excellence: Lessons from Americas Best-Run Companies*. New York: Harper & Row, 1979, page 39.
3. Zuboff, Shoshana, *In the Age of the Smart Machine*. New York: Basic Books, 1988, page 31.
4. Drucker, Peter F., *Innovation and Entrepreneurship: Practices and Principles*. New York: Harper & Row, 1985, page 90.
5. Zuboff, *In the Age of the Smart Machine,* page 33.
6. Zuboff, *In the Age of the Smart Machine,* page 35.
7. Zuboff, *In the Age of the Smart Machine,* page 41.
8. Zuboff, *In the Age of the Smart Machine,* page 42.
9. Zuboff, *In the Age of the Smart Machine,* page 42.
10. Zuboff, *In the Age of the Smart Machine,* page 42.
11. Zuboff, *In the Age of the Smart Machine,* page 42.
12. Zuboff, *In the Age of the Smart Machine,* page 44.
13. Zuboff, *In the Age of the Smart Machine,* page 45.
14. Zuboff, *In the Age of the Smart Machine,* page 238.

CHAPTER 3

Quality and the Bottom Line

THE QUALITY TREND

The idea of quality in manufacturing has probably undergone more change over the past fifty years than any other in industry. Perhaps the primary reason for this is that "quality" is more of a concept than a well-defined attribute and one that can apply to many different things including products, organizations, companies, and individuals. In fact, the word *quality* has been used interchangeably to describe all of the above.

The interviews Foxboro's research team conducted with senior manufacturing executives revealed that the quality trend in manufacturing has undergone three major transitions in the past half century (see figure 3.1). Most of this emphasis on quality has focused on maintaining and improving specific measurable attributes of products being manufactured. This period had been referred to as the "Quality for Manufacturing" period of the quality revolution. During the late 1980s the focus seemed to turn, at least in the process industries, to a much stronger technology emphasis that was based on statistical process con-

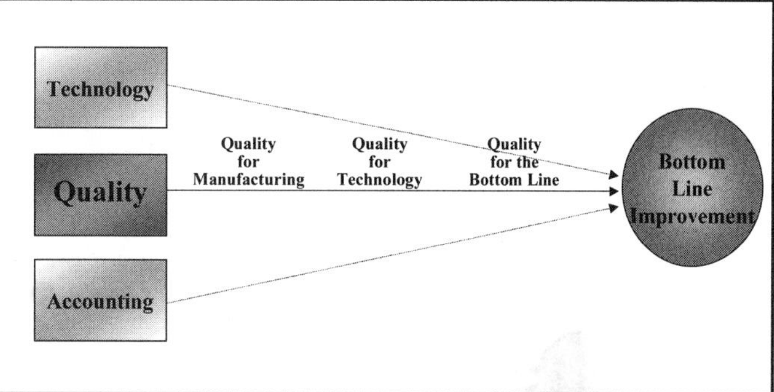

Figure 3.1 The Quality Trend

trol (SPC) tools. This period has been referred to as the "Quality for Technology" phase. Currently, there is a shift underway to a more performance-focused view of quality that is significantly changing the way quality is being addressed in manufacturing operations. It is referred to as the "Quality for the Bottom Line" phase.

The concept of quality has been important for as long as people have made products. But it is a very difficult concept to quantify. If a product exhibits the characteristics that the market needs or wants, it is considered to be of reasonable quality. The more desired characteristics a product exhibits, the higher its quality is deemed to be. Because almost all markets are dynamic, however, with needs and wants that change over time and across segments, this relative definition of quality means manufacturers may be uncertain at any given time that their products have the necessary quality. Thus, one of the first steps in measuring the quality of a product is to develop a comprehensive understanding of the characteristics the market wants.

The quality issue has only grown in importance as globalization has impacted many segments of industry in the past ten years. This globalization has manifested itself in two major ways. First, manufacturers have begun to encounter new competition in market segments where there have traditionally been only established, well-known competitors. New competitors frequently alter the market's perception of what constitutes quality because they introduce the consumer to a wider range of products, influencing their understanding of product characteristics, including quality. Staying on top of dynamic new definitions of product quality requires great effort on the part of manufacturers and, in turn, a corresponding degree of flexibility in their manufacturing processes. Even more challenging, these new product modifications require manufacturers to continuously monitor the market to determine whether their products' characteristics are those consumers require at any given time for any market segment.

The second aspect of the globalization of the marketplace is that manufacturers are being forced to market their products in geographical markets they had not previously participated in.

When they move into these new markets manufacturers often find that their new customers value different product characteristics than customers in their traditional segments. This puts additional pressure on the manufacturer to accommodate the new requirements. Thus, today's globalized market makes it more difficult than ever to manufacture high-quality products on a consistent basis.

QUALITY FOR MANUFACTURING

The first reasonably vigorous approach manufacturers took to ensure product quality has traditionally been referred to as quality assurance (QA). Very simply, QA means checking that the required characteristics of a product were properly built in during manufacture. If the required characteristics are present, the product is released to the marketplace and sold. If they are not, then the product must be reworked, sold at a different grade (and usually at a lower price), or discarded. Whichever of these options is chosen, there is a significant penalty to manufacturers for turning out poor quality products.

This traditional QA approach made sense in the days when manufacturing operations made only one product and market needs seldom changed. But in today's rapidly changing manufacturing environment, the static QA approach is unsatisfactory because a dynamic environment requires dynamic monitoring and control.

Walter A. Shewart pioneered an alternative approach to QA in the 1920s and 1930s while working to improve the quality of telephone equipment at AT&T Bell Laboratories. As noted in chapter 1, Shewart authored the book, *Economic Control of Quality of Manufactured Product,* which presented a structure for controlling product quality by using statistical analysis of the extent by which each manufactured product varied from its specification. Shewart argued that statistics could help manufacturers identify nonrandom and unexpected faults in the manufacturing process that lead to poor quality. Once identified, the faults could be repaired, and the process would make products with the expected quality characteristics. Shewart's work gave birth to a new field that became known as statistical quality control (SQC), a key moment in the quality trend that continues to the present day (see figure 3.2). He successfully implemented an SQC program at Western Electric Company in New York.

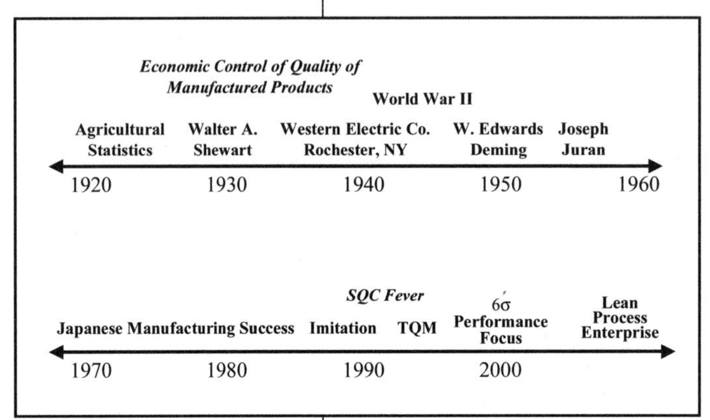

Figure 3.2 The Evolution of Quality Improvement

The basic concept behind SQC is that manufacturing machinery can be combined into a process that, when it is working well, will produce products within an acceptable range of the product's desired characteristics. As long as every machine in the process is working within its designed limits the resulting product will be within specification. It is very important to note that even when all machines are working well, however, the resulting products will not all be identical. There is no such thing as an absolutely perfect manufactured product that exactly matches design specifications. The physical world just doesn't work that way. That is why quality characteristics are often quoted in terms of tolerance bands rather than as absolutes.

For example, if a drill press is part of a manufacturing process for making nuts, two of the key actions of the drilling operation will be to make sure that the hole is in the middle of the piece and that it has the correct diameter. Though the specification for the hole's diameter may be one-quarter inch, if we actually measured the hole in any given nut it could vary, in a very small way, around one-quarter inch. However, such variance is not considered to violate the product's basic specification. The expectation is that if the actual diameters of all the manufactured nuts are mathematically averaged, and assuming that the drill press is working properly, then the average would be one-quarter inch, even though no single part would exactly meet that specification. In other words, if the standard deviation of the holes' diameters from the specification of one-quarter inch is small enough, a large percentage of the parts are considered to be within tolerance.

If a product's specifications are seen as absolute values, then no single manufactured piece will be without error. Machines cannot produce a product, and measuring devices cannot measure any product's characteristics, to such an ideal specification. What this implies is that some degree of error in manufacturing is acceptable, and in fact expected. These errors are normally distributed about the mean, which should equal the specification with a reasonably small standard deviation. This acceptable deviation corresponds to the tolerance. If we measure a product in this way and our measurements exhibit this expected and tolerable degree of error, then the manufacturing process is said to be in statistical control—the machinery is working as well as we can expect. As W. Edwards Deming points out: "statistical control does not imply absence of defective items. It is a state of random variation, in which the limits of the variation are predictable."[1]

When SQC is discussed in terms of manufacturing, the focus of the quality control is on the manufacturing process and based on statistics generated by the indicators of the quality of the product. The product can only be as "good" as the process that manufactures it. If each part of the manufacturing process is working at the expected level, the manufacturing line, as designed, is operating as well as it can be expected to. If this is not good enough, the processing equipment must be changed to introduce a technology that will enable the specification to be met. Because of this focus on the process, the use of SQC tools to provide the information that allows the operators of a manufacturing process to operate at specification and within its limits of variation is known as statistical process control (SPC).

Quality problems in manufacturing arise when the products exhibit unexpected variations from the quality specification. When this occurs, some of the manufactured pieces will be of unacceptable quality because they do not meet the customer's valid requirements. If the quality errors occur randomly, then the existing process equipment is probably working close to its limits.

If the quality errors do not occur randomly, the cause is probably in the process and needs to be isolated and fixed.

Returning to the example of the drill press, suppose that with time the drill bit becomes overheated, causing its shank to bend slightly. The holes that the drill produces will become larger than normal, and the result will be nuts of unacceptable quality. A statistical profile of the resulting parts would show that the distribution is no longer normal, with the same characteristics clustered around the desired specification, and that the average and/or standard deviations are larger than expected. This would lead us to conclude that an unusual situation in the drilling process is causing this unexpected result. Identifying and repairing the cause in the manufacturing process should result in a return to quality production, at least until another unexpected problem occurs.

The statistics, therefore, are used not to identify the existence of variation in the quality characteristics, since variations exist in every manufactured product, but to identify unexpected errors caused by an unusual circumstances in the manufacturing process. If the cause can be identified, the process can be repaired and returned to producing parts with acceptable quality characteristics. Shewart felt that we should use statistics at every stage of the manufacturing process to identify and fix process errors as they occur, thus minimizing the amount of "off-spec" product manufactured. This statistical approach allows manufacturers to control the quality of product as it is made rather than waiting to check until the end of the process, as with QA.

One other key aspect of Shewart's work his realization that it was impractical to measure each manufactured part as it is made, so to make the statistical approach more pragmatic he suggested sampling groups of parts *randomly* and then applying sample statistics to these groups. He also suggested using statistics other than just the average and standard deviation to provide a more complete profile of what was actually happening in the process.

Although Shewart's approach seems to be very clear and extremely practical, manufacturers were very slow to adopt it. Very few plants seem to have used SQC in any systematic way before the 1950s. Perhaps this is because there was such a great demand for manufactured goods and such small supply that product quality was not as much of an issue as product volume. Also, perhaps the dynamics of manufacturing and the marketplace didn't call for much change. Whatever the reason, like so many other novel ideas, Shewart's SQC was not widely implemented in its originator's lifetime.

SQC IN MANUFACTURING

The circumstances in Japan after World War II created an ideal environment for implementing new concepts in manufacturing. Much of the Japanese manufacturing capability had been destroyed, and certainly the Japanese economy was in ruins. Rebuilding Japan was a top priority both within Japan and for other countries, such as the United States.

The United States invested heavily in the rebuilding effort with both financial and human resources. One of the most renowned of these human resources was a statistician by the name of W. Edwards Deming. Deming was a talented scientist who had studied Shewart's SQC method while working with him at Western Electric Company. He found in Japan an ideal test bed for implementing SQC in actual manufacturing environments.

Deming offered this new SQC approach as a way to help Japan reestablish a manufacturing base, and in fact a dominant position in the world. As he said, "I think that I was the only man in 1950 who believed that the Japanese could invade the markets of the world and would within five years."[2] And the rest, as they say, is history.

The implementation of SQC techniques did not go as smoothly as many would have us think. Although the statistical approaches were found to be very revealing, getting manufacturing people to act on this information was not easy. The laborers in these plants did not necessarily see it as part of their job to study statistical charts, and engineering staff tended to be too scant and focused on other issues to do the job on their own. *It became clear that quality was more than a technical issue; it was also a "people" issue and an organizational issue. "Statistics alone cannot assure reduced variation," Deming wrote; "it takes everyone, management and labor working together."*

One approach to getting workers to use the SQC information was to develop teams around each "work cell" in a manufacturing operation and give them responsibility for managing the product characteristics that impacted the product quality—the quality indicators—for their cell's operation. This approach achieved some success and justified the SQC approach, but it did not accomplish all of the desired results.

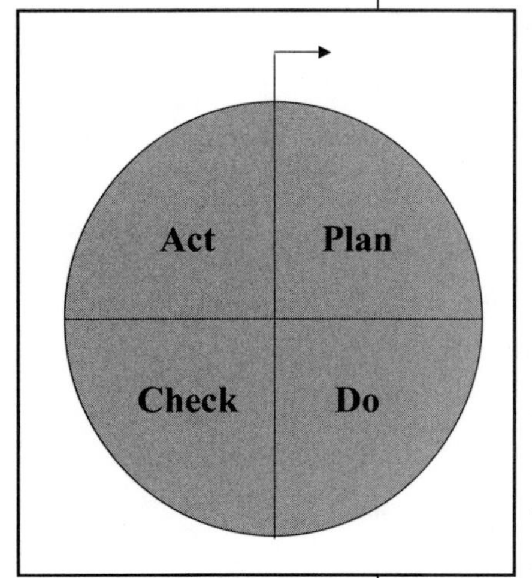

Figure 3.3 PDCA

One of Shewart's concepts was that identifying and eliminating the causes of nonrandom, unexpected error in the manufacturing process "on an ongoing basis" would lead to continuous improvement of quality. He built a simple model based on a cycle for *continuous improvement*, often referred to as P-D-C-A for Plan-Do-Check-Act (figure 3.3). The model suggested that quality improvements would be the inevitable result if the statistics were used to identify process problems and plan for their correction, then execute that plan, check to see if the plan worked, and act on the results. Since the goal is continuous improvement, once the cycle is complete you cycle through it again and again, improving the process, perhaps just a little, each time around.

Deming referred to this P-D-C-A model as the "Shewart Cycle," although most others today refer to it as the "Deming Cycle." In fact, it appears that Deming was the first to implement this approach on any large scale through his work with quality circles. As a result, *continuous improvement* became one of the catchphrases of the quality movement.

This concept of continuous improvement required a significant change in management philosophy, which was dominated at the time by Scientific Management and are still common today. This change included delegating responsibility and authority for managing the "quality indicators" right down to the frontline workers in the manufacturing operation. These are the people with the most pragmatic and detailed understanding of their segment of the operation and are, therefore, in the best position to identify and correct the conditions in the process that cause quality problems. Scientific Management was based on treating the "laborers" as repetitive machines, and not insightful personnel.

Deming was enormously successful in implementing statistical techniques for quality control in manufacturing plants. The basic quality circle approach, however, didn't seem to go far enough to motivate people to work spontaneously toward continuous improvement, as he had hoped. One major reason for this was that delegating the appropriate level of authority to the quality circle required a management approach that was somewhat revolutionary at the time. The "bottom-up" perspective implicit in the idea of quality circles did not spontaneously prompt a corresponding change in management philosophy.

Four years after Deming went to Japan, another pioneer in the quality movement, Joseph Juran, lectured in Japan on *managing for quality improvement.* Juran also had a keen understanding of Shewart's techniques and had also worked with him at Western Electric Company. Rather than focusing on statistics, Juran focused more on the organizational aspects of a manufacturing operation, and in particular the top-down management acceptance and directives that would lead to the continuous improvement of quality and make it an inherent part of the organizational philosophy.

Juran believed that if quality improvement concepts were to reach their full potential, a fundamental change had to occur in manufacturing operations, starting at the highest management levels. He realized that the quality implementation had to occur at the front lines of the operation, that the people on the front lines had to have the culture and structures necessary to do their job, and that there had to be a system of recognition to match. Juran proposed that organizations would manage effectively for quality by allocating the responsibility for quality management to the frontline workers, encouraging teamwork to solve the quality problems, and then recognizing a job well done. As obvious as this approach may seem to us, it was diametrically opposed to tradition. Traditionally, management "ordered" and labor "did," period. To bring labor integrally into the management process was revolutionary.

To prove his theories, Juran set up a total management approach for quality. His approach included an aggressive effort on management's part to plan the quality structure, a structured approach to implementing P-D-C-A, and an effort to change the way people worked to incorporate the quality concepts in day-to-day activities. The structure for managing quality represented an entirely new management and organizational opportunity for a plant or factory. Many operations actually set up two separate organizational charts—one for the traditional functional operation and one for the quality operation, even in reality though the two structures are literally inseparable.

Juran proposed the formation of a quality council at the executive level to plan and administer the quality improvement process. Under the quality council were one or more levels of lead teams who oversaw the quality program in each functional segment of the operation. At the lowest level were the quality teams that actively cycled through the P-D-C-A model to improve quality.

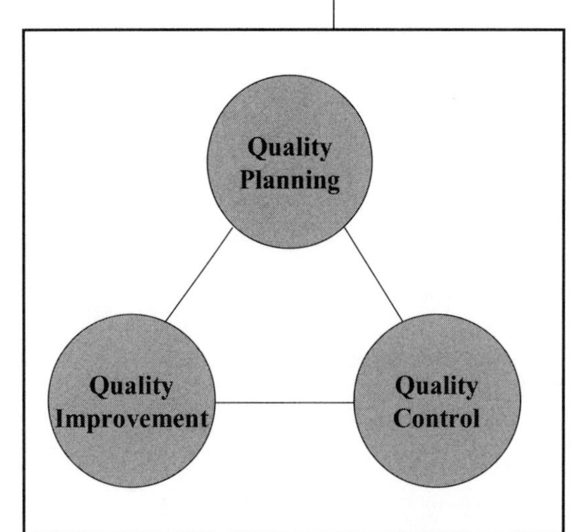

Figure 3.4 Juran Trilogy

The quality teams worked with a structured problem-solving approach to identify and correct quality problems in a well-defined, systematic process. The quality problems were defined as those issues that negatively impact valid customer requirements. As the number of corrected problems rose, so too did the quality of the product and organization, because more valid customer requirements were being met.

After the quality planning was done by the quality council and the quality improvement was implemented by the various teams, the third part of the "Juran Trilogy" (see figure 3.4) was day-to-day quality control. This entailed making P-D-C-A a normal part of everyday work. The result was a change in culture to a true focus on the customer's valid requirements by everyone throughout an organization.

Through this vigorous approach, Juran showed that improved quality involved both effective tools, such as the statistical tools, and a management approach and structure to effectively implement the tools. As James Gagne has noted, Juran "recommends using statistical process control but warns that it can lead to a 'tool oriented' approach."[3] *Moving to high quality is not just a technical issue; it is even more a management issue. Manufacturers who take this view show remarkable results and become global manufacturing leaders.*

Incidentally, it would be a mistake to leave the impression that Dr. Deming only worked with the statistics and did not address the management requirements of a quality program. In fact, he developed many of the same management conclusions as Dr. Juran.

The importance of the quality movement's transition from a statistical approach to a management philosophy cannot be overstated. Today, many companies refer to their quality program as Total Quality Management (TQM) rather than SQC. The management implications of an organization moving to a TQM approach are enormous. The traditional top-down, militaristic management approaches that have been so ingrained in the manufacturing culture are directly opposed to a true TQM approach.

The TQM approach totally redefines the relationship between management and "labor," and in fact, almost eliminates the traditional connotation of labor. Frontline workers must become part of the management team, as must the entire organization, for a TQM structure to be most effective. Each individual must be viewed by the operation, and must view him- or herself, as an important manager of quality improvement.

The concept of quality has changed over the past fifty years. It has evolved from fixed product characteristics to a whole corporate philosophy based on "continuous improvement." Those manufacturers who have taken seriously the total quality improvement approach have moved into an elite category of world-class manufacturers.

QUALITY FOR TECHNOLOGY

Over the past few decades, Japanese companies have moved into dominating positions in a number of manufacturing sectors. As this happened, companies from the other industrialized areas of the world have tried to emulate the Japanese success factors.

Customers with firsthand knowledge of the continuously improving Japanese quality began putting pressure on traditional manufacturers to provide products with the same quality. Immediately, traditional manufacturers wanted to be *seen* as stepping in line with the new and successful Japanese approaches. They wanted to demonstrate to their customers that they were using the same approaches as the "high-quality" competitors.

Quality management, as Juran proposed it, is not overly visible on initial inspection, but the tools of SQC are highly conspicuous. As a result, during the 1980s an epidemic of "SQC fever" started to sweep the industrialized world. Statistical charts started popping up on office walls and corridors across the globe as proof of corporate commitment to continuous improvement.

It is not clear how much of an improvement in quality most manufacturers attained. As a matter of fact, many didn't even appear to be using or even looking at the statistical charts their efforts generated. The stated objective was to implement a quality improvement program, but in many cases attention was paid to window-dressing, demonstrating to customers that a quality program was in place. Implementing a true and comprehensive quality management program did not seem to be much of a consideration.

What happened was exactly what Juran had feared. Manufacturers focused on the *tools*, not on a comprehensive program to continuously improve quality. Perhaps it was this mindset that led Dr. Hans Bujaria, president of Multiface, Inc., to point out in a discussion on SPC that "seven years and billions of dollars later, many users are still waiting for appreciable return on their investment in SPC."[4]

This SQC fever is similar in many respects to the technology-for-its-own-sake mentality that governed the application of computer technology to the control of manufacturing operations, as described earlier. This time the "technology" was based on statistical techniques and tools. Obviously, without a firm commitment to quality improvement, SQC fever was not destined to succeed. Worse yet, the absence of appreciable results tended to prove to the skeptics that it was not really worth investing in SQC and that the quality movement was more of a fad than a trend. The fact is, since most were treating it as a fad their words became a self-fulfilling prophecy.

SQC FEVER IN THE PROCESS INDUSTRIES

The original SQC statistical tools and techniques were developed for discrete parts manufacturing operations. Even Shewart's early work was almost entirely focused on the discrete manufacturing segments. The operations at Western Electric Company, where Shewart originally tested his concepts, manufactured such items as electromechanical switching relays for telephone systems—a classic discrete manufacturing operation.

In discrete manufacturing operations, products tend to be made by machining and assembling pieces and parts. These operations are naturally "discrete" since the same basic functions are performed on parts that are physically separate or distinct. Applying sample statistical techniques in this type of environment is relatively straightforward. The statistics developed for basic SQC were based on discrete manufacturing concepts in which individual samples were randomly selected, measured, and analyzed. Shewart applied random sampling techniques because there were too many parts to effectively measure the whole population.

Perhaps the best way to examine the impact of SQC fever is to investigate their application to the manufacturing segments in which SQC did not originate: the process industries. In the process industries, product is more often produced through combinations of movements of fluids flowing together through vessels and pipes. Intuitively, these manufacturing processes are clearly considerably more continuous in operation. Nevertheless, many engineers who applied SQC techniques to the process industries did so by artificially "cutting" continuous flows into discrete segments, thus pretending that the fluid, non-distinct nature of the process did not exist. This approach has been shown to be technically questionable by a number of statisticians. To demonstrate its shortcomings, let's examine simple examples in which statistical tools are applied to both continuous and discrete processes.

In the last chapter, SQC concepts were introduced through the example of the discrete operation of drilling a hole in the center of a part that was to become a nut. The two quality indicators for this operation were the centering and diameter of the hole. As long as the hole was consistently and reasonably centered and reasonably close to the correct diameter, then from the perspective of this one manufacturing operation the process was under statistical control. Note that centeredness and diameter are truly reflective of the characteristics that the customer values in the product since if the hole does not have the correct diameter the nut will not work as it should, and if it is not centered, the user will have difficulty using the nut. In either case, the nut will not be viewed as having an acceptable level of quality.

Now consider a simple continuous process operation that involves the flow of a liquid through a pipe. Suppose the parameter of interest in this operation is the temperature of the liquid. A temperature sensor is inserted through the pipe into the liquid. To apply the same statistical approach as we applied to the nut would require us to take a random sample of the liquid and develop averages and standard deviations from groups of samples. This has been accomplished traditionally by taking a snapshot of the temperature at different points in time using time-based sampling techniques. Although one could question the randomness of any sample selected using these techniques, it is not necessary to do so for the purposes of this discussion. The averages are calculated from a group of the snapshot temperatures across some number of time samples. Using the same logic Shewart used in developing his SQC method, our process is in statistical control as long as the averages are normally distributed about the desired temperature specification, with no obvious pattern in the distribution.

The heat in the liquid exhibits thermodynamic characteristics and, essentially, physically averages itself across the liquid in the pipe. When the sample is taken and averaged, an average of an average is being calculated. By the Central Limit Theorem of statistics, the average of an average is always normally distributed regardless of the structure of the initial distribution. *The result is that the traditional statistical chart derived from the sampling of a continuous process will not reflect manufacturing process problems in the same way as it would in a discrete manufacturing process.*

Repetitive patterns are a second characteristic that is traditionally looked for in the control charts because it might indicate an abnormal process condition. Most process plants have quite a high level of deterministic control, primarily through the use of feedback controllers. Feedback controllers inevitably introduce oscillation into the process to bring the controlled variable back to the desired set point. Feedback controllers actually force repetitive patterns or oscillations into the variables. Therefore, in process operations even the existence of a repetitive pattern cannot be considered to be giving the same information as it does in discrete operations.

The point of all this is *not* that SQC techniques have no application in process plants—they do. Rather, it is that these statistical techniques and approaches cannot be applied in the same way to these two very different types of manufacturing operations.

Temperature was chosen as the quality indicator in the previous example primarily because it is typical of the type of indicators chosen when SQC tools are applied in process plants. It is important to compare the relative meaning of the selected quality indicators in both the discrete and continuous examples. As was already pointed out, the centeredness and diameter of the hole in the nut directly reflect the customer's true requirements for the product, and continuous improvement along both of these indicators means better quality.

Temperature, however, is a process variable, not a product variable. What does it mean to "continuously improve" temperature? How can temperature be made "better"? This question clearly doesn't have the same meaning as it would for other valid customer requirements. The process variable "temperature" in the liquid stream does not have the same connotation as the quality indicators of "centeredness" and "diameter" do for the nut. The concept of "continuous improvement" has a different meaning in process manufacturing than it does in discrete manufacturing.

Good deterministic decision-making and control in which the corrective action is directly calculable is superior to stochastic decision-making and control in which the corrective action must be inferred from statistical data. The center and diameter of the hole in the nut are not, and typically cannot be, deterministically controlled as the drilling operation is underway because the technology to accomplish this is either not available or too costly. However, most process variables can be, and are, deterministically controlled. If a variable is deterministically controlled using feedback or feedforward control techniques, does it make sense to also apply stochastic analysis to the variable? Control experts agree — it does not.

The real issue is abandoning the unproductive focus on the tools designed to help people continuously improve quality and focusing instead on the desired outcomes. For the past decade, the technology-driven approach to quality has been almost as illogical as the technology-driven approach to automation discussed in chapter 2. Industry must not abandon the quality movement; rather, it must work to make it effective.

QUALITY FOR THE BOTTOM LINE

The significant question now becomes: "What are the true quality indicators in a process plant?" In other words, how *should* the concept of continuous improvement be applied in a process environment? Since process plants utilize a significantly higher level of direct deterministic control than discrete factories, the answer involves a further level of abstraction in the concept of quality and valid requirements.

To develop effective quality indicators for process operations, the Foxboro team interviewed dozens of operations managers and production managers from various segments of the process industries. After we explained the dilemma of identifying, selecting and measuring quality indicators, we asked them a very simple question, "What in your plant would you most want to improve on a continuous basis?" The answer was equally straightforward: "plant performance." In fact, this response was so basic that it was almost discarded as unhelpful, but at second glance, it proved to be much more significant. As Peter Drucker wrote in his book *Innovation in Entrepreneurship*, "an innovation, to be effective, has to be simple. It is vitally important that we do not overlook the obvious."[5]

The next question on our survey was, "What is your plant's performance now?" Typically, the answer was that *they really didn't know*—operations and production manager were not even sure how to measure the key parameters of performance in their plants. Using the same basic thought process employed by Shewart, Deming, and Juran, it is painfully apparent that one cannot continuously improve on performance when the basic parameters of performance have not been established!

As basic as the concept of continuous improvement is to plant performance, it is also the most significant issue in process operations, and it is becoming more significant in the non-process sectors as well as their level of sophistication increases. The quality indicators of plant performance are, in fact, very similar to traditional quality indicators in discrete manufacturing in that they represent true value to both the manufacturer and the customer—even if at a higher level of abstraction than traditional quality indicators. These performance quality indicators hold the key to the success of quality control in process environments and in any performance-based operation.

The shift in focus to a performance-based approach to quality improvement is well aligned with some of the more intensive quality improvement programs that have become popular in major organizations in recent years. One good example is the 6σ (or "Six Sigma") program, which has been driven through a number of major global companies such as Motorola, General Electric, Allied Signal, and Invensys. The 6σ program is based on the same basic tool set as the TQM programs it replaces but has a much more aggressive performance focus. The name 6σ itself denotes this shift in emphasis over previous approaches. The Greek letter σ or sigma represents standard deviations from a target in statistical analysis. A standard deviation of 3σ means that 99.9 percent of all parts will be within an acceptable range, resulting in 1,349 defects per million parts made. Most good manufacturing processes have been around the 3σ level. A standard deviation of 6σ means that the defect rate has been reduced to 3.4 parts per million. This is a very aggressive target, but adopting it clearly demonstrated the seriousness of the effort to gain performance improvement through these programs. It is also important to note that most of the jargon surrounding programs such as 6σ was still oriented very much toward discrete statistics and discrete parts.

When Dr. Juran structured the TQM approach, one of his guiding principles was that performance improvement teams should be charged with selecting their own quality indicators, developing a method for measuring them, and then developing and executing a strategy for continuously improving them. His idea was that the people on the "front lines" can identify the issues that need correcting much more effectively than the managers who are removed from the details of the process. His concept was that each time a team or individual goes through a P-D-C-A cycle a little improvement would be realized. If enough people and teams went through enough improvement cycles the total improvement would become quite significant. In essence, his approach was to just trust the teams or people responsible for improvement to select the right quality indicators and improvement strategies.

This laid-back approach to quality improvement did not sit well with many manufacturing executives, especially given the tough global marketplace where exceptional levels of performance improvement were required for survival. These executives essentially agreed that a quality improvement program was a good thing, but they wanted to ensure that the quality indicators and improvement projects selected would have an immediate and strong impact on the economic performance of their manufacturing operation.

The traditional approach to TQM was based on the idea that people basically understand how the organization needs them to perform and will therefore select the most appropriate quality indicators to drive company performance. This is not the case in most manufacturing operations. As Robert Kaplan and David Norton pointed out in their *Harvard Business Review* article "Measuring Corporate Performance," "Executives also realize that traditional financial accounting measures like return-on-investment and earnings-per-share can give misleading signals for continuous performance improvement and innovation."[6] Without valid performance information and direction for the quality improvement teams, many executives have questioned the validity of the quality improvement projects and quality indicators that were selected.

This debate prompted an analysis of the difference between quality indicators and performance measures. Performance measures are quantitative metrics designed to indicate how every group, process, and person in an organization is performing with respect to the organization's strategic objectives. Quality indicators are measures of the quality of delivery of parts, products, or services from an individual or group to their direct customer (figure 3.5). The customer can either be internal or external. The differences between performance measures and quality indicators have been much discussed, but lack of valid performance measures has rendered much of the discussion moot. Dr. Juran appears to have had a clear idea of the difference, but he assumed that organizations would have effective performance measurement systems in place. He assumed that if teams or individuals

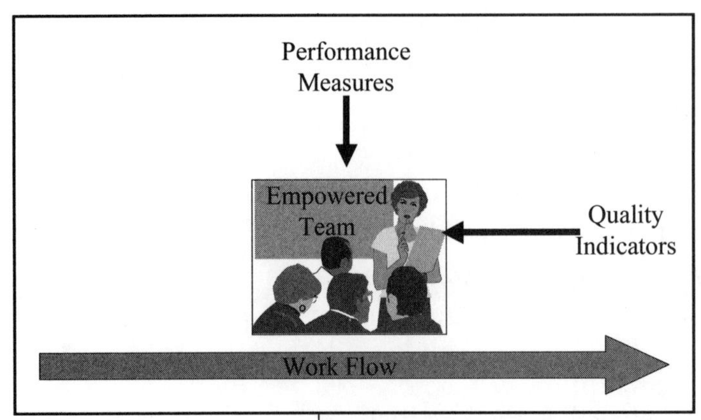

Figure 3.5 Quality Indicators or Performance Measures

truly understood how their behaviors impacted the organization's strategic objectives and understood their job within the organization then they would naturally select appropriate quality indicators and improvement projects. In theory, this seems quite reasonable, but in practice the lack of effective performance measures created a significant gap between expectations and perceived results. With a poor performance measurement system, even if the quality improvement teams were accomplishing a great deal of good for the organization, it would be tough to quantify and document the value.

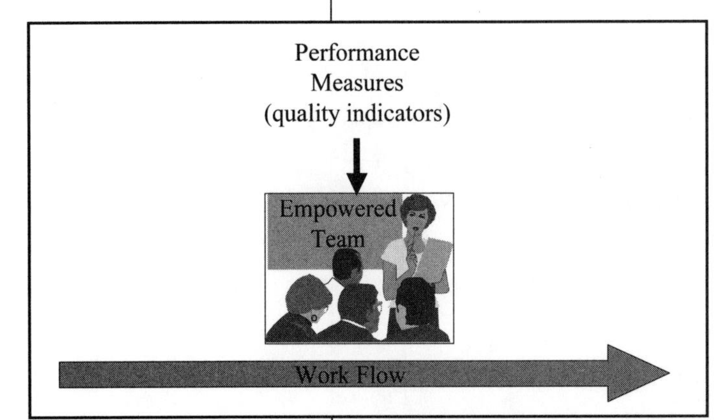

Figure 3.6 Quality Indicators and Performance Measures Converge

These debates over quality indicators and performance measures coincided with the trend toward performance-based quality initiatives, such as 6σ programs. This combination of drivers led to a convergence of performance measures and quality indicators (figure 3.6). In discrete manufacturing environments, this convergence caused considerable confusion and upheaval because these firms' traditional approach to quality indicators was often totally independent of performance measurement considerations.

In the process industries, this convergence injected some logic back into the quality improvement programs. Process manufacturing environments had traditionally had great difficulty with the concepts behind quality indicators, but they had a much stronger historical focus on economic performance. Even though most process manufacturing operations did not have effective performance measurement systems in place, the idea of using performance measures as the quality indicators was a natural fit. All that was missing was an effective performance measurement system with well-defined performance measures for each process unit in the manufacturing plant. In addition, converting the quality improvement program into an individualized, day-to-day program rather than just a quality team approach as Dr. Juran recommended requires that the performance measures be available in a time frame that was appropriate to the individuals' jobs or tasks. In process operations, that time frame is typically measured in minutes or seconds because the operators are overseeing the production of products in process units that can change that quickly. Providing the performance measures daily or weekly would not correspond to the work being done and would limit the effectiveness of a continuous improvement process.

The quality movement has come a long way over the past decade. Many very positive results have been realized through the focus on continuous improvement of quality throughout manufacturing operations. The movement began in discrete manufacturing operations and has enjoyed the most success in discrete environments. The recent drive toward performance-based quality programs has led to a convergence of quality indicators and performance measures. This has provided great incentives in process manufacturing environments to reinvigorate their continuous improvement initiatives. This new convergent approach will drive bottom-line quality and economic performance improvements. The only thing missing in most plants is a real-time performance measurement system that will provide the new class of performance-based quality indicators to measure and drive continuous improvement.

THE CONVERGENCE OF LEAN MANUFACTURING AND QUALITY IMPROVEMENT

Since the 1990 publication of *The Machine that Changed the World* by James P. Womack, Daniel T. Jones, and Daniel Roos, the concepts of lean manufacturing and the lean enterprise have gained a great deal of attention in the manufacturing sector. The book chronicles the success Toyota and several other automobile manufactures realized by changing the perspective of their manufacturing operations from mass production to lean production. The mass production perspective encourages the manufacturing operation to make as much product in a given time as possible while lean production adopts a demand-pull perspective toward manufacturing. Once this change in perspective occurred, many traditional paradigms in manufacturing that had once been unassailable suddenly collapsed. The results were phenomenal. The quality of the products made in these lean manufacturing environments continually improved while the real costs of manufacturing went down. Unfortunately, most of the successful examples of lean techniques realizing positive results have been in the automobile or other discrete manufacturing environments. However, experts in lean concepts intuitively believe that the lean approach should and will work in other business areas and environments.

LEAN MANUFACTURING CONCEPTS

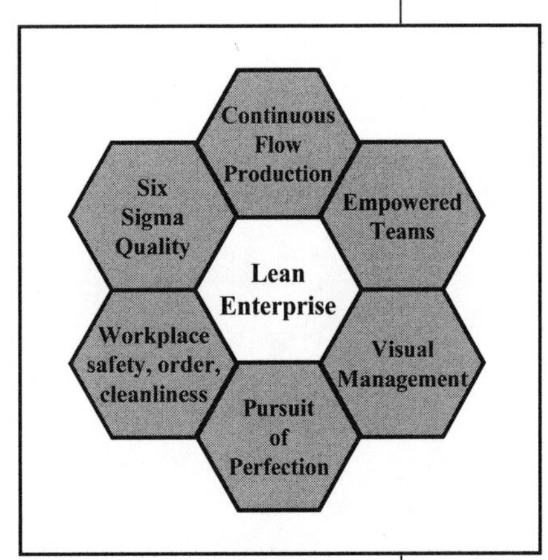

Figure 3.7 Lean Enterprise

Six basic principles form the foundation of the lean enterprise approach as it is described in *Lean Transformations* by Bruce A. Henderson and Jorge L. Larco (see figure 3.7). Systematically applying these principles has led to significant improvements in discrete manufacturing environments. A brief summary of each of the principles follows:

1. *Continuous Flow Production* – This principle aims at converting a traditional isolated work cell approach to production to one that follows a more ongoing or uninterrupted flow. This principle applies to the production operations within an organization as well as across the supply chain.

2. *Six Sigma Quality* – This principle is representative of the drive to achieve zero defects in the products produced by a manufacturing process. Six sigma is a statistical value that implies that 99.9996 percent of all products produced

are within acceptable quality limits. The Six Sigma Quality initiative is a performance-based extension of the Total Quality Management (TQM) approaches introduced by Deming and Juran.

3. *Empowered Teams* – As part of the TQM approach, both Deming and Juran encouraged organizations to create and use quality teams that are empowered to manage the continuous improvement in quality indicators they are responsible for. In recent years, the concept of empowered teams has been extended into more of a general management philosophy for creating improvement in many areas.

4. *Visual Management* – This principle involves presenting the key performance indicators to the empowered teams as well as to all levels of management so as to gain an ongoing assessment of the improvement being realized. The visual management approach encourages the use of easy-to-understand graphics to present this data.

5. *Workplace Safety, Order, Cleanliness* – This principle is focused on the concepts that a well-organized work area can use to encourage improvement. It is based both on the idea of providing an environment that demonstrates to employees that they are valued members of the organization and of promoting orderliness to encourage excellent work habits.

6. *Pursuit of Perfection* – This principle tends to be an overriding principle of all lean operations: they should continually strive for perfection. Though it is clearly an unattainable objective, striving for it helps to drive the continuous improvements that are required to be world class.

Reasonable reflection on these six principles usually leads to ready acceptance. They amount to a collection of common sense ideals. Unfortunately, common sense often has to be systematically applied in order to be applied at all.

LEAN PROCESS PLANTS

The primary principles behind the lean enterprise concepts, as with the principles behind Total Quality Management, had their genesis in discrete manufacturing operations. There is a very tight correlation between these principles and the conditions that have evolved in assembly-line or similar operations. Though there are few examples of formal lean production programs being implemented in process plants, the basic principles definitely apply. The following is a short overview of how each of the lean principles may apply in process manufacturing environments:

1. *Continuous Flow Production* – Most pure process manufacturing environments are continuous flow by nature and design. Batch process environments can be made more continuous by reducing cycle time and time-between-cycles as much as possible. This is similar to the reduction in switchover time in discrete environments. If the time-between-cycles in batch process environments can be minimized, the traditional trade-off between production and flexibility can be significantly reduced. Developing continuous flow through the supply chain in a process environment is a very challenging task. With lean techniques in discrete manufacturing operations, the workflow is regulated by the takt time of the operation. Takt is a Japanese work meaning "the beat to which to adjust the pace of your manufacturing operation."[7] Considering the concept of a takt time in what is already a continuous flow operation is difficult. The key is to treat the entire supply chain like a single continuous flow operation by providing common performance measures across the various organizations within the supply chain. This will enable each component in the chain to manage the flow of production to the downstream group to optimize production and cost.

2. *Six Sigma Quality* – Applying the principles of quality management in process plants has been a challenge ever since Walter Shewart introduced statistical control in the late 1920s. The statistical control tools were designed

around sample statistics in discrete component environments. Process manufacturers have been struggling to apply them in continuous process environments ever since. Part of the problem in both discrete and continuous process environments has been defining the quality indicator to be measured. In discrete operations, quality indicators such as *defects per millions of parts* made are quite reasonable, but how does such a measure translate into process environments? Additionally, Juran encouraged organizations to separate the performance measures and quality indicators of an empowered individual or team.

As TQM concepts made the transition to a more aggressive performance-oriented approach through the introduction of structured, performance-based TQM approaches like 6σ, the distinction between performance measures and quality indicators began to blur. This blurring was actually beneficial to the application of 6σ tools in process plants because, if performance measures were available in real time, the quality indicators for plant performance would be readily available. The tools for continuous improvement in process environments are different than the traditional statistical control tools and include such tools as advanced process control, multivariable predictive control, and process optimization. Nevertheless, the objective is identical. The availability of real-time plant performance measures will make possible the effective application of 6σ concepts in process plants.

3. *Empowered Teams* – The teams process plants need to generate continuous improvement include process engineers, control engineers, maintenance personnel, operations personnel, and production personnel. The best way to empower any team is to provide it with the toolkit and clear objectives and measures for improvement necessary to realize the improvement. The real-time performance measures provide this for the teams, and the performance targets provide the clear objective functions.

4. *Visual Management* – Visualizing plant performance in process plants has been a problem since the earliest days of process manufacturing. Performance dashboard displays at the system workstations that present the performance measures in real time are finally making visual management in process plants a reality. Visual management works. Process manufacturers can finally take advantage of this high-powered communication and performance-enabling approach.

5. *Workplace Safety, Order, Cleanliness* – In process plants, unlike discrete manufacturing operations, the operator is often removed from the process and located in a control room. Control room environments are typically cleaner and more orderly than plant environments, but most control rooms could be significantly improved. One of the key ways to improve the workplace environment in process plants is to incorporate more electronic technology to make it easier for operators to access the information they need, with less clutter. Process automation systems are designed to enable this.

6. *Pursuit of Perfection* – Facilitating the pursuit of perfection in process plants is the essence of continuous economic and quality performance improvement. To pursue perfection and drive continuous improvement, there must be an understood target and a metric system that defines improvement. The real-time performance measures, associated performance CRT dashboard displays, and embedded performance targets help process manufacturers to define perfection and drive toward perfection. Perfection is no longer an abstract ideal in process plants. It is a well-defined set of objectives that plant personnel can work toward.

Lean production and lean manufacturing have been limited to discrete manufacturing operations. The availability of real-time performance measures, performance dashboards, continuous improvement tools, and advanced automation system technology can combine to make lean manufacturing approaches available to process manufacturers.

NOTES

1. Gagne, James, "Quality Performance Means More at Dow." Midland, MI: Dow Chemical U.S.A., page 3.
2. Dobyns, Lloyd, "Ed Demings Wants Big Changes, and He Wants Them Fast." *Smithsonian,* August 1990, page 77.
3. Gagne, "Quality Performance Means More," page 5.
4. Bhote, Kevin R., "America's Quality Health Diagnosis: Strong Heart, Weak Head." *Management Review* (American Management Association), May 1989, page 36.
5. Drucker, Peter, *Innovation and Entrepreneurship: Practices and Principles*. New York: Harper & Row, 1985, page 135.
6. Kaplan, Robert S., and David P. Norton, "The Balanced Scorecard – Measures that Drive Performance," *Harvard Business Review on Measuring Corporate Performance*. Boston: Harvard Business School Publishing, 1998.
7. Henderson, Bruce A., and Jorge L. Larco, *Lean Transformation*. Richmond, VA: Oaklea Press, 1999, page 58.

CHAPTER 4

Cost Accounting and the Bottom Line

THE COST ACCOUNTING TREND

The cost accounting trend is somewhat different than the technology and quality trends in that its origin is more tightly linked to the financial operations within manufacturing companies rather than their manufacturing operations. This was also a much more difficult trend for our research team to gain an understanding of because the team members are automation focused and had much less interest in and knowledge of accounting principles and techniques.

At the beginning of the research program, the executives being interviewed discussed cost accounting as a critical success factor in gaining the benefit of automation. Nevertheless, the research team did not grasp the significance of this factor, perhaps because team members felt that it was obvious that proving the economic value of automation or any other program would require an effective accounting approach. Perhaps the team just assumed that the accounting systems companies currently used possessed all the information required. What the team missed early on was that the executives being interviewed were trying to point out that an effective accounting approach did not yet exist to effectively capture the economic value resulting from automation or any other manufacturing process improvement.

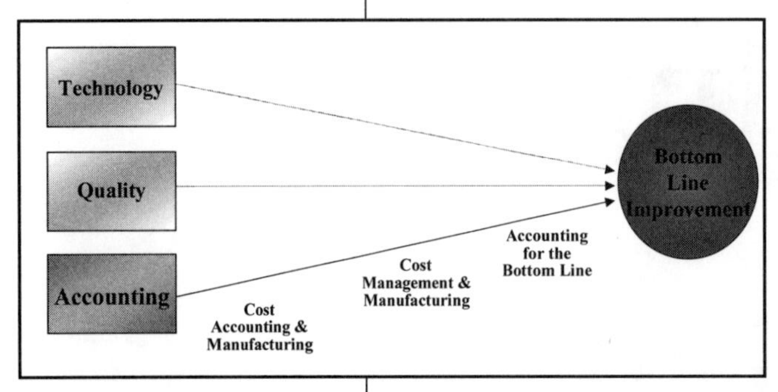

Figure 4.1 The Cost Accounting Trend

As did the technology and quality trends, the cost accounting trend has evolved through three major phases over time and is still evolving (figure 4.1). As the research project began, traditional cost accounting systems had been in place and operating in one form or another for decades without any appreciable modification. Some of the executives interviewed seemed to feel that these systems were merely a fact of life that had to be dealt with. Many expressed serious concern over the lack of effective and timely data. We refer to this initial traditional phase of the accounting trend as the "Cost Accounting & Manufacturing" period.

Very shortly after the research team began to really understand that traditional cost accounting had serious limitations when applied to manufacturing operations, new approaches to accounting in manufacturing environments designed to overcome these limitations were gaining recognition and acceptance. These new approaches were initially labeled "cost management systems" or "activity-based costing systems." We refer to the period of transition from traditional cost accounting to these new accounting systems as the "Cost Management & Manufacturing" period. It became very clear soon after the team recognized these new approaches that many of the executives interviewed were driving their accounting systems in this direction.

As the early cost management and activity-based costing systems started to come on line, they seemed to offer significant benefits over the traditional cost accounting systems. However, serious limitations in these approaches and significant implementation difficulties were soon identified. This has led to a second transition that is currently taking shape in many manufacturing companies, a transition that should lead into a new concept and approach for accounting in manufacturing operations. We refer to this emerging period as the "Accounting for the Bottom Line" phase.

COST ACCOUNTING AND MANUFACTURING

Corporations are in business, for the most part, to make a profit. The way corporations measure whether they are profitable is through cost accounting systems. Years ago, these systems were comprised of hundreds of bookkeepers and accountants manually performing calculations. Today, computers do the bulk of the work. Other than that, the basic structure of cost accounting systems has changed very little over the years.

The executives we interviewed expressed great concern over the usefulness of the information they could gain from their cost accounting systems. They accepted that the cost accounting systems in their operations were necessary for reporting required financial data but did not feel that they provided much data that was relevant to their manufacturing operations. They appeared to believe that "most cost accounting systems are historical, oriented toward financial reporting, and inadequate to measure operational performance."[1]

Cost accounting systems evolved very little between about 1900 through the 1960s. The primary objective of these systems was to report to management and the financial community on the financial performance of companies in a way that was consistent across all companies and operations. This consistency of reporting was necessary to provide effective communication and comparisons of the financial performance of different companies and operations. Over this period, standard cost accounting principles and practices became very well defined and accepted.

From a manufacturing perspective, cost accounting systems had very little impact on the day-to-day operations. This may seem odd since a good percentage of the direct costs of most manufacturing companies occur within their manufacturing operations. But the consensus was that "the main purpose of the old cost-accounting system [was] to help the financial department monitor operations and value inventory."[2] In other words, the primary customers of cost accounting systems were the finance department and not the manufacturing operation, and the outputs from these systems were focused to meet the needs of this constituency.

Despite the strong, widespread emphasis on manufacturing performance since 1970, cost accounting systems, oddly enough, did not evolve to provide performance-based information for manufacturing operations. This lack of progress was closely associated with the technology trend we discussed in chapter 2. As digital computers gained acceptance for the management of business information management during the 1960s and 1970s, one of the first initiatives of many corporations was to automate their cost accounting systems. This made reasonable sense since previously these systems were very manpower-intensive and extremely costly.

In the early years of the computer movement, there were very few professionals with computer programming expertise and no standard software packages. As a result, companies typically hired college graduates trained in technical fields such as mathematics or physics and invested in training them in programming. Programmers with any degree of experience were usually promoted to the role of systems analyst and became responsible for designing the computer program-based systems, such as the cost accounting system.

Computerized cost accounting systems were generally made up of a large number of separate programs, each designed to perform a subset of the entire cost accounting process. Systems analysts would assign the responsibility for creating the programs to several different programmers. The system analyst's job was to ensure that all the programs of the cost accounting system would effectively work together to produce the desired outputs. Structured programming techniques had not yet been developed for the early programmers, and in any case the programmers had been trained to design programs to optimize scarce computer resources, such as memory space or computer time. As a result, the programs were very difficult for programmers who had not written them to understand. To make matters worse, the average tenure of the new programmers in any company was short because once they had even a little experience they could earn more money by joining a different company. Most companies' cost accounting systems were therefore comprised of many unstructured programs written by several different programmers who had since left the company. The new programmers hired to replace them had great difficulty analyzing and modifying the programs, and computerized cost accounting systems went years with only minimal modification. The unwillingness and inability of companies to modify and expand their cost accounting systems actually became one of the main reasons cost accounting did not change appreciably from the 1960s through the 1980s.

The way in which manufacturing performance was measured through traditional cost accounting systems was fairly universal across all industries (see figure 4.2). The quantity of each product manufactured over a given period was determined and entered into the cost accounting system. Cost accountants then determined the total cost to make the product and run the plant over this period, and then calculated the *cost-per-unit-product-made*. This statistic, cost-per-unit-product-made for each product or product line manufactured, has been the essential method for measuring manufacturing performance since the very earliest days of manufacturing.

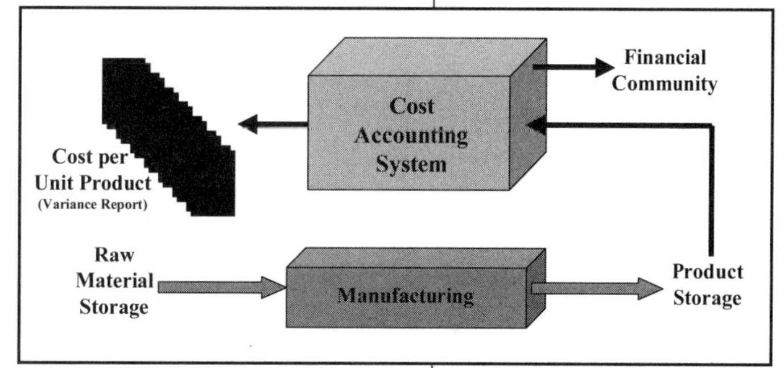

Figure 4.2 Cost Accounting and Manufacturing Performance

This information has typically been transmitted back to manufacturing management through biweekly or monthly variance reports. The idea behind a variance report is that the accountants would establish cost targets, or "standard costs," for each product made at the beginning of the fiscal year based on their most optimistic prediction. The cost-per-unit-product calculated each month by the cost accounting system was compared to the standard cost, and from this the variance from standard was calculated. If the calculated cost was less than the standard cost, the manufacturing operation was determined to be doing a good job. If the actual cost was below the standard cost on an ongoing basis, the manufacturing manager for that operation might receive a healthy raise or even a promotion. If the calculated cost was greater than the standard cost, the manufacturing manager had to be prepared to do some explaining.

The primary objective of the manufacturing operations within these corporations has been to make product at the required volume and the lowest possible cost. Clearly, this is a key element in any cost-based structure. Years ago, the bulk of the costs incurred by most corporations came during the manufacture of the product. Since overall cost, including manufacturing cost, was the primary focus of the cost accounting systems, the variance reports became the basis for calculating the cost-based manufacturing performance measures. Measures of an individual's or organization's performance are referred to as *performance measures*. In the case of traditional cost accounting, the performance measures have been cost-per-unit-product-made.

Since the financial community measures the performance of a corporation by its profits, the traditional cost accounting approach for measuring manufacturing performance would seem to be the most reasonable accounting method. Indeed, it was reasonable seventy years ago when most manufacturing operations made few products, typically on separate manufacturing lines, and when most of the cost of an operation was directly assigned to the products made (direct costs). Overhead costs, such as general administration costs, comprised only a small percentage of any product's total cost. Also, historically manufacturing operations were viewed as necessary, but not strategic, operations within the corporation. Gaining competitive advantage through flexible manufacturing or repeatable quality was not considered to be of prime importance. Within this context, the primary driver of manufacturing operations should probably have been cost, and the cost-per-unit-product-made metric for measuring performance was probably reasonable.

The situation in manufacturing operations today is vastly different. Many operations make multiple products, even on the same manufacturing lines. Moreover, "Overhead costs have risen dramatically and now exceed direct labor costs... It is not uncommon to find that direct touch labor accounts for only 8 – 12 percent of total cost at many factories."[3] Finally, in today's difficult competitive environment, manufacturing is often viewed as strategic and essential to the execution of the company's strategic plan.

To compensate for the large percentage of an operation's total cost that cannot be directly assigned to products, cost accountants today use allocation algorithms that apportion the overhead costs that cannot be directly assigned to the products manufactured (figure 4.3). "An allocation process assigns costs that have been accumulated in a cost pool to a particular management reporting objective."[4] These allocation objectives may be based on any number of factors, such as market demand for each product, volume of product sold, or even price. "Overhead (indirect labor, depreciation, energy costs, maintenance, supplies etc.) is lumped into cost pools and then absorbed into the cost of the products according to some factor – usually the number of 'standard' direct labor hours in the product."[5]

$$\textbf{Cost per Unit Product} = \frac{\textbf{Direct Costs + Allocated Costs}}{\textbf{Quantity of Product Produced}}$$

Figure 4.3 Traditional Product Costing

The way the allocation is accomplished is not the key factor. The important point is that manufacturing managers are being measured by the fraction, *cost-per-unit-product-made*. They have little or no control over the value of the numerator (cost) of this fraction. The largest percentage of the costs indicated by the numerator is allocated to costs incurred by the operation that the manufacturing manager can do little to impact. Even if a manufacturing manager does an excellent job controlling cost, this might impact the numerator of his or her performance metric by as little as one tenth of one percent, and because of allocation algorithms it could impact another manufacturing managers' performance measure by as much, or even more.

As an example, I remember sitting in a meeting of manufacturing executives a few years ago as a manager told a story involving the allocation of cost in his facility. It seems that one of the products manufactured in his facility had been determined to be a real loser. The cost accounting system was consistently showing it to have negative margins. Since some valued customers really liked the product, the corporation decided it could not just pull it off the market.

A team was formed to study the problem and determine how to phase out the product without negatively impacting the customer base. After a few months of team meetings, the product began to show positive margins in the variance report. Corporate management was pleased and congratulated the team on their good work, but the team was astonished because they knew they hadn't done anything that would have caused this change. They decided to wait another month to see if the improvement in margins was just a fluke, or if something had really changed. The next month the cost-per-unit for this product on the variance report was even lower. The team investigated to see what had changed to cause this favorable result. They checked raw material costs, energy costs, and every other cost they could think of, but they could not find any appreciable factor that would have made this difference. They checked for changes in product volume and selling price, but still found no appreciable change. Finally, when they were about to abandon their search, one team member talked to a company cost accountant and learned that the overhead allocation algorithms for the plant had been changed to make the allocations more representative. The change in allocation approach was the cause of the performance improvement for the product. The manufacturing manager and the manufacturing operation for this product had absolutely nothing to do with the favorable trend in measured performance. They appeared to be performing considerably better because less overhead cost was being allocated to their operation. I can't help but wonder if the manager got promoted as a result.

"The calculation and aggregate reporting of overhead by traditional cost accounting systems has hindered management's understanding of what comprises overhead costs and the cost implications of non-productive activities."[6] Also, as a result of the overhead allocation and the increasing percentage of cost that is overhead, the numerator of the cost-per-unit-product fraction generated by traditional cost accounting systems for manufacturing performance is almost totally beyond the control of manufacturing personnel. Therefore, the only part of this measure that is within their control is the denominator: the amount of product made over a given period. Manufacturing managers who are interested in performing well according to the performance measures of traditional cost accounting systems will always make as much product as possible at all times regardless of the manufacturing strategy. In doing this, they can elevate the denominator and correspondingly lower the overall value of the fraction. As a result, the actual manufacturing strategy for most companies in the traditionally industrialized areas of the world over the past few decades has been to make as much product as possible regardless of market demand.

In recent years there have emerged several manufacturing strategies and action plans that have been the center of much attention in the manufacturing world. One of the most significant of these is just-in-time (JIT) manufacturing, introduced in chapter 1. JIT is a manufacturing action plan with wide-ranging implications, one of the most important of which is that manufacturing operations are only supposed to manufacture to meet existing demand. That is, they should only make the amount of product that the market needs at any given point in time. This allows inventories to be significantly reduced, resulting in a corresponding reduction in the cost to store, maintain, and manage the inventory. This all makes perfect sense, unless you happen to be a manufacturing manager who is interested in driving the performance measure *cost-per-unit-product-made* down because you know this is how your performance will really be judged. Notice that the implementation of JIT essentially takes the denominator of this fraction out of the control of manufacturing management, and that JIT is essentially opposed to traditional, volume-based manufacturing strategies. In a JIT manufacturing environment, a manufacturing manager who performs in line with the JIT action plan can be accused of real performance problems because the only controllable aspect of performance, product volume, has been taken away.

A story a manufacturing manager shared with me about JIT helps to illustrate this point. A JIT program was initiated in this manager's corporation. The success of the program was to be measured by the JIT team representatives walking around and recording the amount of in-process and product inventory that each of the manufacturing operations had and then developing a monthly JIT report. Corporate management, claiming they were committed to JIT, would reprimand the managers of the operations that were not meeting the JIT objectives. But the primary measure of the manufacturing performance remained cost-per-unit-product-made.

The JIT team members walked around with their clipboards for the first two weeks of every month recording inventory levels. It took the final two weeks for them to compile their reports, which they did in offices some distance away from the manufacturing operation. The manufacturing manager told me that for the first two weeks of every month he had the best JIT operation the company had ever seen. But as soon as the door closed behind the JIT folks, production was cranked up as high as possible. The manufacturing team hid product in closets, cars, and wherever else they could store it. In this way, the operation was seen as being good at JIT and still measured up to the cost accounting-based performance measures. Something is obviously wrong when the manufacturing people have to work *around* the system to be successful.

It may be worth investigating the failure point of the traditional cost accounting systems with respect to manufacturing performance. One principle that appeared to be shared by all of the manufacturing executives we interviewed is that people will perform to their measures. This is true regardless of what the measures are. If the measures are appropriate, they will drive appropriate behaviors. If the measures are bad, they will drive inappropriate behaviors. As in the JIT example, manufacturing organizations have performed to those measures even when they are not appropriate. For cost accounting to drive the correct operational behaviors, the performance measures generated must be aligned with the operation's objectives.

As programs like JIT have been attempted, the failure of the traditional cost accounting-based performance measures has become more obvious. Manufacturing management has had very little control over these measures except to drive production volume up as much as possible. This has led to the development of huge manufacturing storage facilities such as warehouses or tank farms, which actually add cost to the overall operation. Since that cost is typically allocated so no specific product or product line is affected more than any other, this strategy makes sense within the manufacturing measures of performance.

As the globalization of the marketplace accelerated and the competitive environment for many manufacturing companies became very difficult, many companies have tried to make manufacturing more of a strategic weapon. Unfortunately, the cost-per-unit-product measures have often opposed the strategic initiatives that would align the manufacturing operations with the strategic plan. For example, if manufacturing flexibility is seen as a key strategic differentiator, the drive to increase the denominator of the performance measure would tend to hinder manufacturing flexibility. If a trade-off must be made between a stated program and a measure, people will typically make sure they perform to the measure. If the measures of performance do not represent the strategy, the strategy will probably fail. "Manufacturing managers are being asked to make important decisions in spite of available cost accounting information, not because it is relevant."[7] Clearly, cost accounting systems and the resulting manufacturing performance measurement systems were in critical need of improvement.

COST MANAGEMENT SYSTEMS

As the weaknesses of the traditional cost accounting systems began to become apparent, initiatives to rectify these systems were attempted. One of the first involved trying to revise the allocation algorithms for overhead costs to make them better reflect the manufacturing operations. Manufacturers quickly found out that this approach was futile. "It is an illusion that changing the basis of allocation will solve all cost management problems."[8] Perhaps the most visible and widely accepted approach to modifying cost accounting was based on the concept of "activity-based costing" and went under the general banner of "cost management systems" (CMS).

The primary objective of the CMS approach is to "fix" cost accounting so that more accurate cost data can be determined. The basis for the fix is that too much overhead cost has to be allocated to provide the cost information with any accuracy. Management could "no longer accept an environment where cost accounting contributes to overhead rates so high as to obscure true product costs."[9] The theory behind this movement was that if more of the costs in an operation could be directly assigned to a product or product line, less would have to be allocated (figure 4.4). The resulting cost information should more accurately reflect the true cost of the operation and the products.

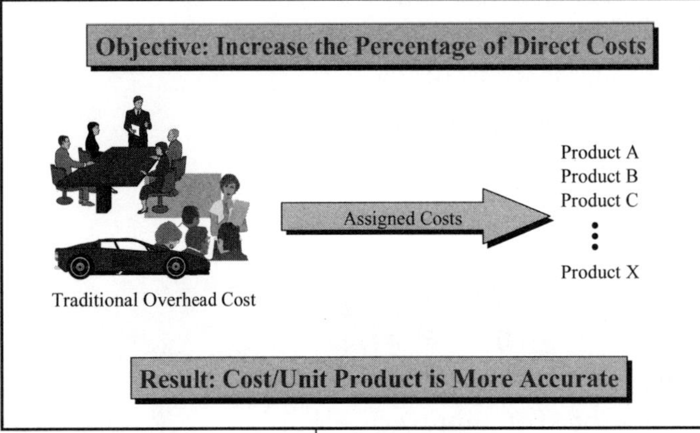

Figure 4.4 Cost Management Systems — Activity-based Costing

From a cost accounting perspective, there is considerable merit to the idea and approach of activity-based costing since it should lead to more accurate cost profiles for the products and any other cost-based activities throughout an operation.

The way to directly assign more of the overhead costs in the operation is to keep much more accurate records on how traditional overhead costs are incurred and then determine a way to directly assign some of these costs to products or product lines. Traditionally, the cost of functions such as product development, marketing, and sales has been considered as overhead. Any costs incurred by these functions have gone into the general overhead pool for allocation. In reality, a large percentage of the time in these functions is spent on a particular product or product line, and the cost associated with this time can and should be directly assigned. This is a key aspect of activity-based costing. Costing systems based on this approach can significantly reduce the amount of overhead that has to be allocated and can result in much more accurate cost information.

As cost management systems gained in popularity during the 1980s and 1990s, many who had started to realize that the cost accounting-based performance measurement systems were flawed saw activity-based costing as the solution. Obviously, more accurate cost information should result in more accurate calculations of cost-per-unit-product-made. In theory, the resulting cost-based performance measures might be more helpful in providing manufacturing with the information it requires managing performance on an ongoing basis.

One shortcoming of this point of view is that, regardless of how accurate the statistic is, manufacturing management and operations personnel have absolutely no control over the traditional overhead costs, even if they are directly assigned instead of allocated. A manufacturing manager has little, if any, direct control over product development costs, sales and marketing costs, or general administrative costs. The fact that these costs are more accurately assigned to products or product lines does not mean that they provide effective manufacturing performance measurements.

A second issue, stemming from the tough, competitive global environment, is that manufacturing has began to be considered as an important competitive weapon in the execution of company strategies. In a cost-based strategy, cost-per-unit-product-made may be a reasonable metric, but if the corporate strategy is based on flexibility, quality, or production, cost-based metrics fall far short. Japanese manufacturers have moved into a strong position in manufacturing over the past twenty years. Certainly, they measure the performance of their manufacturing operations: "Japanese companies put much more emphasis on measuring the non-financial aspects of factory performance."[10] It is critically important that we learn from their success and not fall victim to the quick fix. Cost management systems represent a significant advance in the accuracy of cost accounting, which is important for financial management, but manufacturing performance measurement requirements and cost accounting requirements may not be as tightly coupled as they have been in the past. Cost accounting systems "employ performance measurements that often conflict with strategic manufacturing objectives, and they cannot adequately evaluate the importance of non-financial measures such as quality, throughput and flexibility."[11] Moreover, "Managers need up-to-date, concise information formatted to assist them in making the right decisions."[12] Perhaps it is time to look for a whole new performance measurement system, one that will work to move companies to world-class manufacturing through world-class performance.

ACCOUNTING FOR THE BOTTOM LINE

It should be fairly obvious at this point that in order for an accounting system to have strong bottom-line impact on manufacturing operations it must produce effective and accurate manufacturing performance measures. "Financial performance measures" should "indicate whether the company's strategy, implementation and execution are contributing to bottom-line improvement."[13] Traditionally, corporate and manufacturing performance measures were viewed as byproducts of cost accounting systems. They were treated as almost an afterthought. For manufacturers to compete in the competitive global environment, this must change. "Enhanced competitiveness depends on starting from scratch and asking: 'given our strategy, what are the most important measures of performance?'"[14]

The effectiveness of any company's performance measurement system must be treated as a critical factor to the success of the execution of its competitive strategy. One critical success factor to developing a good manufacturing performance measurement system is to make certain the system is measuring the performance parameters that really matter. "Many companies do not institute a measurement system that supports the company's strategy."[15] A performance measurement system that is not aligned with the company's strategy is a formula for failure. It is very important that the performance measures represent the corporate strategy and the resulting strategy of the manufacturing operation. The only reasonable way to establish this linkage is to clearly define the corporate strategy and ensure that the manufacturing performance measures are derived directly from it: "Performance measures should help establish congruence between organizational and company objectives."[16]

As we discussed in chapter 1, Dr. Thomas Vollman, a professor at the International Institute for Management Development, developed a simple model to help visualize and systematize this activity that is sometimes referred to as the "Vollman Triangle" (figure 4.5). It can be very useful in analyzing and developing effective corporate performance measurement systems so we will describe it again here. At the top of this triangular model is the plant's manufacturing strategy. The strategy has to be clearly represented and communicated to all key people in the operation if their performance is to support the strategy. Out of the strategy should come an action plan indicating how the strategy should be implemented. The performance measures should be developed to represent and measure the success of the action plan: "Performance measures should provide the link between the activities of the business and the strategic planning process."[17] If the performance measures reflect the strategy and are available at all levels of the organization, the company's personnel will tend to perform to the measures and, thus, the plant's manufacturing strategy will be effectively executed.

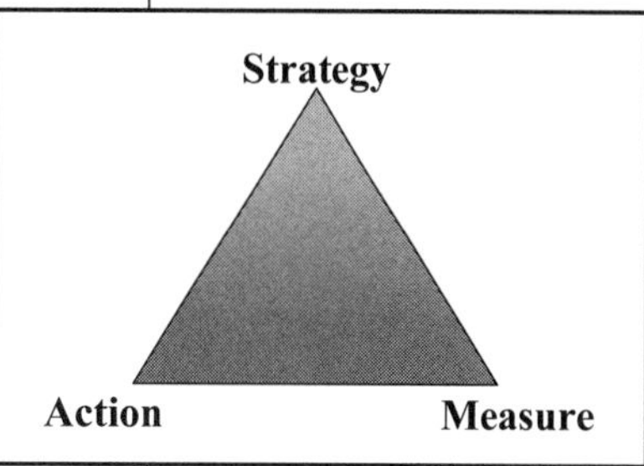

Figure 4.5 "Vollmann Triangle"

"Cost is only one aspect of performance ... greater emphasis is given to providing non-financial information reporting for decision support."[18] Japanese manufacturers have demonstrated an understanding of the concepts portrayed by Vollman's model. If the action plan requires financial measures, then financial measures should be developed. If the action plan requires non-financial measures, then non-financial measures should be developed. To effectively execute most manufacturing strategies usually requires a mix of financial and non-financial measures. If the right measures are utilized, the action plan will be implemented successfully because the employees will understand how they are measured and will act appropriately. As Harvard's Peter Drucker points out, "the scarcest resources in any organization are performing people."[19] Perhaps this is because most of the people in most organizations just don't know how to perform because they have no way of determining how their actions impact performance.

The Vollman Triangle can be viewed as a continuous improvement model for manufacturing strategy similar to the "Plan-Do-Check-Act" model of Total Quality Management. It represents a cycle that starts with the strategy. From the strategy an action plan is developed, and performance measures should be constructed to measure the effectiveness of the action plan. Once the performance measures are under control, or market conditions dictate that a new strategy is required, the Vollmann Triangle cycle can be utilized over again, resulting in continual improvement of manufacturing performance in line with the competitive strategy.

In traditional industrial plants, manufacturing strategies and action plans are often developed, but they are measured and evaluated via the traditional cost-per-unit-product-made measures. If these measures happen to be accurate and match the action plan, there is some chance that the correct strategy will be implemented. But if the measures do not match the action plan, which is most often the case, the probability of the strategy succeeding is almost zero. Many manufacturers have seen their manufacturing strategies fail for exactly this reason.

It is critically important that the measures of manufacturing performance precisely match the plant's manufacturing strategy and action plan so that the operation, and every person in it, is taking informed actions based on fulfilling the strategy. "There needs to be an explicit link between the performance of individuals with the performance of the company as a whole."[20] Any "measurement is meant to enable us to do, to take purposeful action based on knowledge rather than opinion or guesswork."[21] The purposeful action desired in any manufacturing facility should be directed at making the strategy work. Thus, the plant's performance measures must match the plant's manufacturing strategy.

One of the classic challenges to executing competitive strategy has been gaining alignment throughout an entire company. A company can have the most effective competitive strategy in a market, but if its management team cannot determine how to take the high-level corporate strategy down to an actionable and measurable plan at all levels in the operation, the strategy will not be effectively implemented. Once again, Dr. Vollmann has helped to provide a fairly straightforward methodology for driving the strategy through an organization. As we discussed in the foreword to this book, this structured methodology is referred to as a "Vollmann Decomposition" (figure 4.6). The concept behind the Vollmann Decomposition is rather straightforward, but experience has demonstrated that it requires significant corporate discipline and strong leadership to execute the decomposition down through all appropriate levels of the organization.

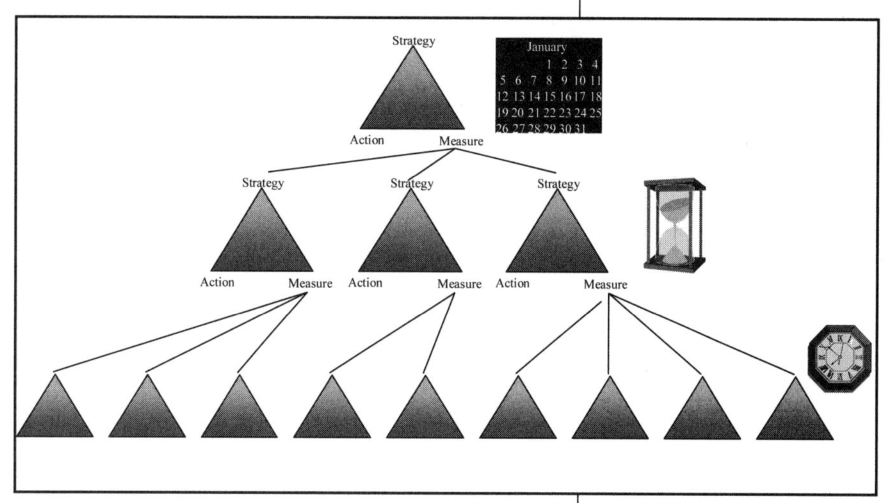

Figure 4.6 "Vollmann Decomposition"

The Vollmann Decomposition is accomplished by top-down repetition of the Vollmann Triangle model. The executive team of most high-performing companies typically develops or adjusts the corporate strategy (see figure 4.7) on a regular, although infrequent basis. For most companies this happens about once per year. The corporate strategy should include an overall strategy statement that provides the short-term vision for the company. Experience has demonstrated that good strategy statements should encapsulate the vision of the company, define the company's current and desired states (if different), and define the customer value the company expects to provide. Since the strategy

- **Strategy Statement**
  - **Visionary**
  - **Current and desired state**
  - **Customer value proposition**
  - **Corporate value proposition**
  - **Short (2 to 3 sentences)**
  - **Simple and Clear**
- **Strategic Objectives**
  - **Decompose strategy statement**
  - **Reflect movement to desired state**
  - **Simple**
  - **Few (< 7 total)**
  - **Short (1 sentence)**
  - **Customer focused**
  - **Actionable**

Strategy Statement

Strategic Objectives
1.
2.
3.
4.
5.
6.

Figure 4.7 Strategy

statement should typically be rather concise, it is important to also develop a set of strategic objectives that define the specific goals and objectives of the company for the coming year. The objectives should provide an added level of specificity to the corporate strategy. Good objectives must be simple, understandable, clearly stated, customer focused, and actionable. Strategic planning teams seem to have a tendency to want to develop a large number of strategic objectives. This can be a big mistake because too many objectives can result in presenting a confused strategy. As a rule of thumb, keeping the objectives to seven or fewer can help strategies stay clear and simple.

- **Action Plan Structure**
  - **Objective Action Supports**
  - **Action Statement**
    - **Tied to strategic objectives**
    - **Identifies "gap" (as is - desired)**
    - **Measurable**
  - **Responsibility**
  - **Schedule**
    - **Starting time**
    - **Completion time**
    - **Progress milestones (if necessary)**
  - **Progress**

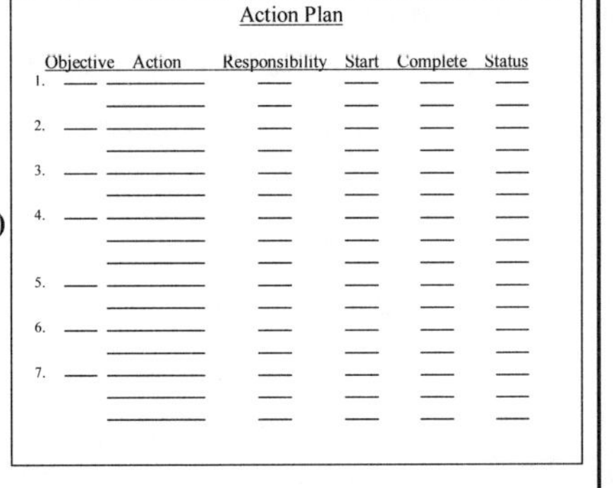

Figure 4.8 Action Planning

As we have seen, the second step in the Vollmann Triangle model is to develop a corporate action plan (figure 4.8). A good action plan is comprised of a set of measurable corporate action statements that are derived from the strategic objectives. The action plan should define a person or organization responsible for executing each action statement, and it should define a schedule. The action plan can be a very good management tool if it is treated as a "living document" and the status of each action step is updated throughout the action planning time period.

The final step is to identify the corporate performance measures from the action statement. Since each action statement should be constructed to be measurable, if the action statements are developed correctly the performance measures should be obvious. Sometimes multiple action statements are developed to drive the same performance measures. In these situations, the number of corporate performance measures could be fewer than the number of action statements. Also, not all of the performance measures defined in the corporate plan are necessarily impacted by all operations in the organization. Therefore, operations, such as manufacturing, may focus their plans on only a subset of the corporate performance measures.

The strategy, action plan, and performance measures at the corporate level are typically too abstract and too far removed from any department or employee to be used effectively in guiding their day-to-day-performance. A structured decomposition is required to bring these down to the level where they drive department and individual performance.

The sample decomposition model in figure 4.6 shows the decomposition structure for a corporation comprised of three major operating divisions. The first level of decomposition is from the corporate performance measures to the divisional level. Each divisional executive team should evaluate the measures of performance defined by the corporate strategy and develop its strategy for driving those measures in the desired direction. From the divisional strategy should come a divisional action plan and divisional performance measures. The divisional performance measures will not necessarily be the same as the corporate performance measures, but should be indicative of the contribution the division can make to the corporate strategy. The key critical component of this structured approach is that, if it is done correctly, as the divisional metrics start to move in the correct direction the corporate measures should also start to improve, by design. If this does not happen, the result could be a seriously dysfunctional organization with different operations moving in very different strategic directions.

If a particular division is comprised of four manufacturing plants, each plant management team should analyze the divisional performance measures and develop a strategy for driving those measures in the correct direction. From this plant-level strategy should come the plant action plan and plant performance measures. This level of the decomposition is sometimes referred to as the "plant manufacturing strategy" since the ultimate reason for the existence of each plant is to manufacture products. The decomposition should continue down through the operation until each group and person within the operation has a clear understanding of how their performance is measured by the company. Following a structured decomposition such as this will ensure that all people and departments within the company are working toward the same objectives. It will also ensure that as each individual in the organization drives his or her performance measures in the desired direction, the corporation, as a whole, will benefit.

One interesting and important element of the decomposition that is critical to the successful implementation of strategy is the timing entailed by each layer in the decomposition. As was pointed out, the *corporate* strategy, action plan, and measures are typically derived annually due to external business planning issues. The *divisional* level analysis should be done according to the competitive and market dynamics that impact the competitive environment of that division. This may be more frequent than the corporate plan. In many of today's markets this could be necessary on a semiannual or even quarterly basis. The *plant-level* analysis must typically be done more frequently than the divisional level analysis because of the dynamics of the local markets and local supply and demand changes. Plant-level plans may need to be executed as frequently as quarterly or even monthly to be effective. As the decomposition continues down through the plant the time-constant requirements for the performance measurement process become shorter and shorter. Fortunately, major strategy changes are not typically done below the plant or divisional level, but action planning and performance measurement systems must continue to be decomposed and must be able to handle very short time constants to be effective. In process manufacturing operations the time constants can be as short as minutes or even seconds. This type of time requirement is totally out of the scope of traditional cost accounting systems or even the more recent cost management systems.

In reviewing the Vollmann Decomposition approach with the executives taking part in the research project, our project team found that many of them had started to drive a very similar strategy decomposition in their operations, although few were as rigorous and highly structured as the Vollmann Decomposition. The team also found that the decomposition tended to terminate at the plant management level for virtually all of the plants investigated. The plant management teams we interviewed tended to understand what their performance measures were, but most of the engineering, maintenance, and operations personnel within the plant did not. The research team ended up referring to this phenomenon as the "automation economic gap" (figure 4.9).

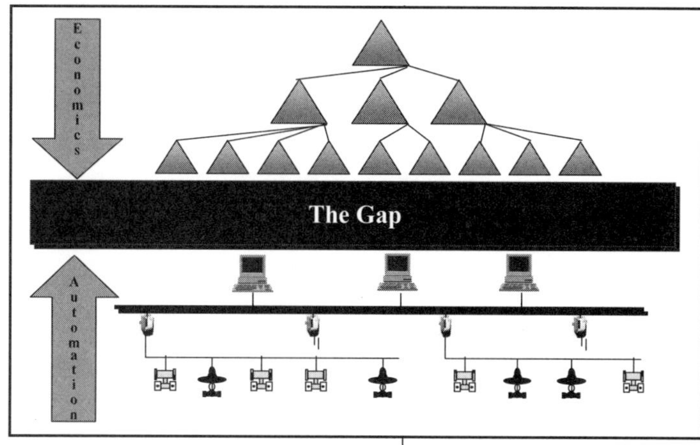

Figure 4.9 The Automation Economic Gap

Although most plant automation systems were installed to help the corporation manage the performance of its manufacturing operations, the performance measures did not tend to make it far enough down to even determine the performance impact of the automation investment. A number of plant operation and engineering teams were interviewed throughout the research process. They were often asked if they knew what their individual or group performance measures were. Most seemed to believe they knew what their measures were, but when we asked them to identify them the measures they identified were seldom aligned with the management-level performance measures.

One example of this was particularly interesting. At one chemical plant of a major global chemical company, a plant operations and engineering team was interviewed and found to be using very advanced automation techniques, including multivariable predictive control and nonlinear optimization. When asked what the objective function of their nonlinear optimization was, they responded that they were optimizing to reduce the manufacturing cost within the defined constraints of the process. This seemed quite reasonable until the plant manager was later interviewed and told us that the primary plant objective derived from the corporate strategy was to increase production. The engineering team was doing an absolutely excellent job of optimizing the process to the wrong performance measure. The more successful they were, the less successful the plant management team was at meeting its primary objective. This is a clear example of the gap in action.

It is interesting to note that the concepts surrounding the Vollmann Decomposition approach and activity-based costing are fairly well aligned. Identifying the activities for activity-based costing systems is often done through a similar decomposition approach: "Functions can be decomposed into processes that represent the ongoing set of activities."[22] In a manufacturing environment, the "manufacturing processes would be considered activities."[23] Done correctly, the financial performance measures would be the key activity-based accounting values at every level right down to the process units. One key difference between traditional activity-based costing and the financial performance measures derived from a Vollmann Decomposition is that the economic measures may include more than just cost. A Vollmann Decomposition may also result in nonfinancial measures. In fact, driving the performance measurement system from the corporate strategy right down to the manufacturing process units would lead to a new type of cost management system approach. This new approach may be best referred to as activity-based accounting, and it must be capable of being accomplished in real time as dictated by the time constant of the processes being managed.

DYNAMIC PERFORMANCE MEASURES

Suppose accurate performance measures could be developed that truly matched the plant's manufacturing action plan. Suppose a system were put in place such that these measurements were generated and reported to manufacturing personnel on the same basis as traditional variance reports, let's say at the end of every other week. By the time the manufacturing people knew what their performance *was*, the product they were making at the time the measure was taken would probably already be at their customer's site. There is not much that they will be able to do with the performance information to improve the product or the operation. Also, even if the performance measures indicate that the manufacturing operation had done a good job, they probably will not even remember what they were doing back when the measures were taken that caused them to do well, and they will have great difficulty replicating their success. Performance measures have to be much more timely than traditional variance reports to be truly effective and ultimately useful.

"To be most useful ... the frequency of reported information should follow the cycle of the production process being measured."[24] The manufacturing performance measures must be presented to plant personnel in a time frame that will allow them to respond to the information effectively. They must be real time: "Performance data should be cost-effective, available and timely. They should be reported in a timely basis and in a format that aides decision making."[25] If the people and operations who are actually manufacturing the product can get the measures of their performance *as the product is being produced*, they will be able to perform better and as a result the entire plant will also perform better. As completely obvious as this may seem, very few manufacturing operations have put this into practice.

Measurements of manufacturing performance that reflect the plant's manufacturing strategy and are reported to the manufacturing personnel on a real-time basis are called *dynamic performance measures*. The word *dynamic* has two meanings in this phrase. First, it is intended to reflect the real-time aspects of the measures. Second, it reflects the fact that as the manufacturing strategy changes the values being measured has to change to reflect the new strategy.

The computer-based technologies currently available in manufacturing operations provide the promise that dynamic performance measures are technically feasible. As Harvard's Peter Drucker points out, "computers have now made activity-based cost accounting possible; without them it would be practically impossible."[26] Computer-based control systems are often directly connected to the manufacturing process via process sensors that measure the process flows, levels, temperatures, pressures, and many other process variables. In process plants there are typically hundreds and even thousands of sensors measuring dozens of different types of process values and making these values available to the control system on an immediate, or real-time, basis. To construct dynamic performance measures requires making use of this sensor-based data to model the performance measures of the plant on a real-time basis, right in the computerized process control system. This approach has been tried in numerous process plants, and it has been determined that if there is enough process instrumentation to enable the control of the process there is almost always enough to enable the modeling of the performance measures. In this way, real-time performance measures can be made to represent the performance of every process unit and the entire manufacturing process on a continuing real-time basis, at exactly the same time as the product is being made. Since the measures are based on "live" process values, there is no need to construct a model of the plant as has been attempted so many times in the past. The information provided by the sensor-based measures represents how the plant is actually operating. As the plant dynamics change, and they will, the measures will immediately reflect the new dynamics.

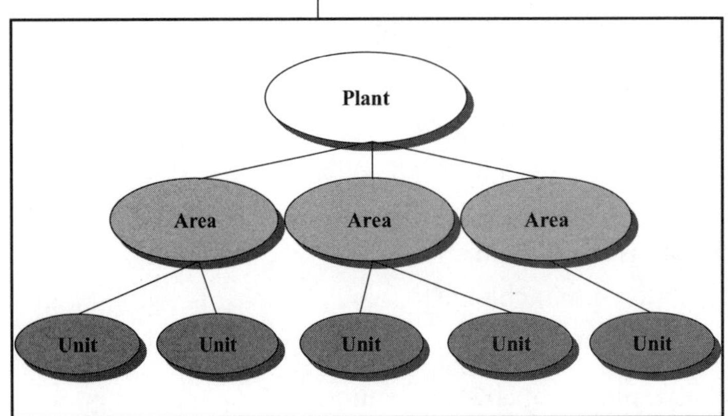

Figure 4.10 Process Plant Decomposition

The specific performance measures must directly relate to the job responsibilities of the people to whom they are presented. "Each performance measure should be totally within the accountability of the person or group performing the activity to be measured."[27] In most plants, the first-level operators have responsibility for a subset of the entire plant operation. The operators are typically assigned an area of responsibility according to the natural decomposition of a plant into areas, and areas into units, and so on (figure 4.10).

The plant performance measures must likewise decompose into the same areas of responsibility. The decomposition must be done in a top-down manner from the plant-level strategies right to the unit-level performance measures (figure 4.11). Once the decomposition has been accomplished down to the unit level, the unit-level performance measures can be modeled off of the process instrumentation data for that unit. Since this data is typically available at the natural time constants of the processes being measured, the models can run in the automation system in real time. When these measures are modeled in real time they are dynamic performance measures. This approach will make the first-level dynamic performance measurements available to the plant operators for that section of the plant in the same time frame in which they perform their operations activities. This type of information in this type of time frame is exactly what the plant operators require to understand how to perform better in their jobs.

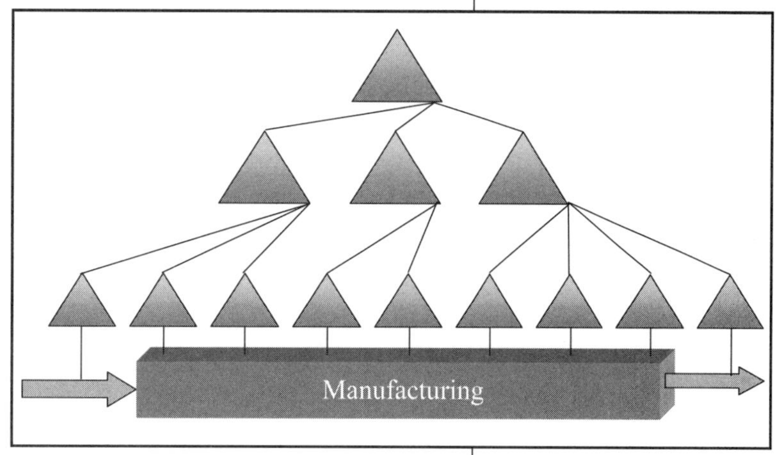

Figure 4.11 Plant Level Vollmann Decomposition

A hierarchy of dynamic performance measures can then be developed by "recomposing" the plant performance measurement hierarchy according to the (figure 4.12). The implementation of dynamic performance organizational structure and domain of responsibility of every individual in the organization measures must be performed bottom-up, in the reverse direction as the initial decomposition analysis: "Operations need both bottom-up ... and top-down know-how about

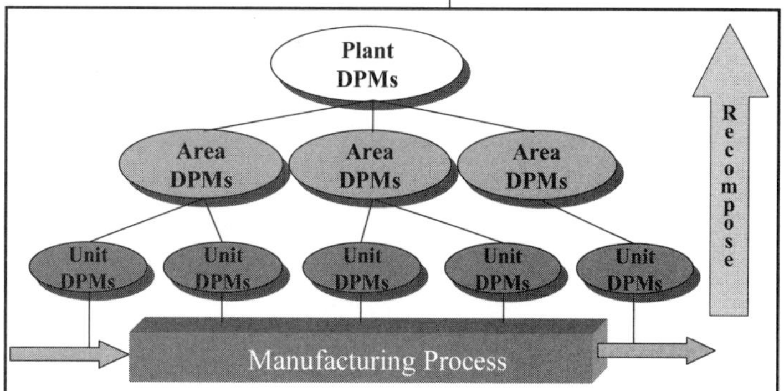

Figure 4.12 Dynamic Performance Measure Hierarchy

the plant."[28] Area supervisors, for example, will have responsibility for the dynamic performance measures that are aggregated from the dynamic performance measures of each of the operators reporting to them. This hierarchy of performance measures provides needed bidirectional access. The lower-level performance measures provide the input for calculating the higher level. It also provides the measures of performance to the people who have the responsibility and authority to manage the performance at each level of the operation in a totally consistent manner across the entire organization. As the dynamic performance measures are aggregated, the time constants of higher-level measures will typically be longer than the measures at the lower level. This characteristic matches the time constants of the job functions being performed.

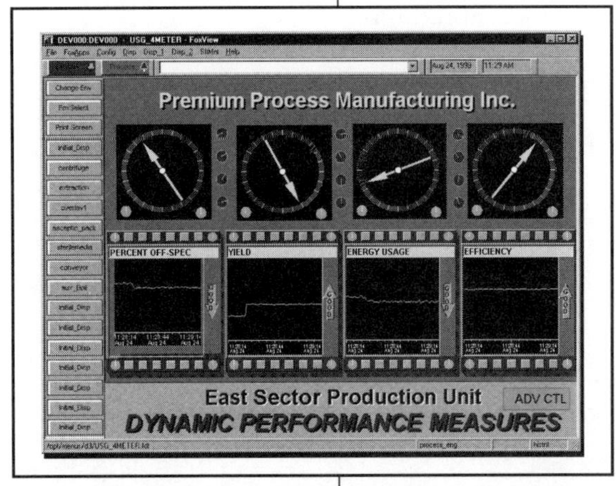

Figure 4.13 Performance Dashboard

The visual presentation of dynamic performance measures to all manufacturing personnel should be simple and straightforward. For example, the graphic approach commonly used to present statistical data in a SQC environment is well accepted and simple, and traditional process gauges have proved to be easy for process operations personnel to deal with. Figure 4.13 shows an example of a display that incorporates both of these concepts. This display is referred to as a "performance dashboard" because it encompasses all the information necessary for an operator to drive the performance of the plant in a form similar to an automobile or airplane dashboard. The modeled performance measures can be displayed along with their current performance targets and an indicator that shows which direction of movement means good performance. Notice that the dashboard in the figure only displays the value of four performance measures for the operator. "It is critical to keep the number of measures down so that there is no ambiguity about where people should be focusing their energy."[29] Also, very often the performance measures have a tendency to "fight" each other. As an action causes one measure to improve, the same action may cause another measure to move in the wrong direction. "Performance measures must, therefore, be prioritized according to those strategic success factors deemed most critical."[30] In the dashboard display shown in figure 4.13, the measures are prioritized by their placement on the display: left to right with the left-most measures being higher priority.

The performance dashboards can follow the dynamic performance measures up through the organization (figure 4.14). The dashboard displays can be relatively consistent with the measures that are displayed to reflect the area of operational responsibility at each organizational node. This structure ensures that all personnel in the operation are working in concert and serves to encourage teamwork because all people at each operational node are working to improve the same measures.

Dynamic performance measures represent the application of an important new management *tool* that has been missing from traditional manufacturing systems—a tool to implement real-time activity-based accounting models right down to the plant floor. It expands the scope of traditional cost accounting well beyond just financial measures to a new and more wide-ranging concept of performance measures. Using this real-time activity-base accounting, management can finally use the accounting system as a real-time decision support system throughout their organizations. This new approach can help every person in the organization work to drive the organization's bottom line in concert with its strategic plan. This changes the focus of accounting from after-the-fact cost accounting to accounting for the bottom line.

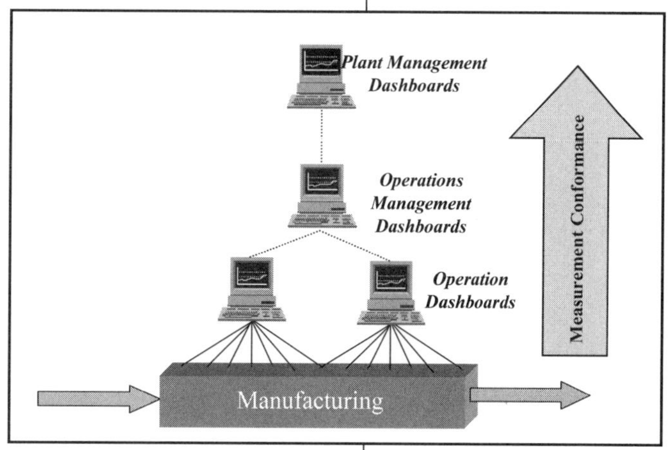

Figure 4.14 The Performance Measure Hierarchy

This real-time activity-based accounting tool in itself can become the *first step* toward world-class manufacturing but, by itself, is clearly not enough. Management must view this as merely a tool that allows them to fill the economic-automation gap that has been a major barrier to performance improvement and strategy deployment for decades. The advent of real-time activity-based accounting models merely provides them with a facility to help them drive performance and execute competitive strategy. Management must be willing to exert leadership in using this new technology to drive their manufacturing operations to new, world-class levels of operation and performance.

DYNAMIC PERFORMANCE MEASURES AND ACCOUNTING

This chapter has explained how Dynamic Performance Measures (DPMs) were a direct result of the evolution of process requirements that traditional cost accounting systems did not address. Once DPMs were installed and operating in process plants, the logical question then became whether and how they should associate with and link into these traditional cost accounting systems. Many DPMs are direct models of the cost or profit associated with a unit or area of a manufacturing process. One of the objectives of accounting systems is to resolve cost and profit issues, but the first DPMs to be installed were typically totally independent of their plant's accounting systems.

This separation of the DPMs from the accounting systems was initially not of much concern. The DPMs were being installed to generate information that had not traditionally been available, so making them available to accounting systems did not seem imperative, or even particularly desirable. DPMs were typically installed in the section of a process plant in which a drive was underway to perform activities that would improve economic performance. The DPMs provided the operations, engineering, and maintenance personnel with direct, real-time feedback on the performance of their section of the plant, and also allowed them to determine the economic impact of specific improvement activities. The problem was that when an activity had been completed and its economic value determined, the plant's management would often ask the finance department to provide confirmation of the economic improvement. The only tool the finance department had to do this was the installed accounting system. As explained earlier in this book, the cost accounting system lacked the information necessary to confirm the improvement. If the finance department could not confirm the improvement, as far as plant management was concerned the improvement had not occurred.

This is not a new issue. Over many years, for example, a considerable amount of intellectual power has been focused on optimizing the operations of oil refineries. Sophisticated techniques such as linear and nonlinear programming have been effectively optimized both refinery planning and scheduling and unit operations. I clearly recall how very impressed I was with the optimization team of one refining company as they discussed and presented the advanced methods they were using in their daily operations to ensure that their refineries were operating at the highest potential. During our conversation, there was quite a bit of debate among the refinery team members, which demonstrated their unprecedented level of commitment, experience, and capability in applying highly sophisticated mathematical models to their operations. As the discussions concluded, I expressed my admiration for the team and asked if they knew how much economic value they had generated for their company in the past year. They responded with a value well over $20 million. I then asked if their CFO agreed with this assessment. After a bit of a pause, the leader of the group stated that over the many years he had worked for the company the finance department had never been able to confirm the value the group had generated. He went on to say that management would not accept as accurate the value the optimization team said it was generating in the refineries unless the finance department could confirm it.

In another case, DPMs were installed in a section of a pharmaceutical plant. Plant personnel determined baseline values for the measures by operating the DPMs so the operators weren't aware of them and periodically storing the DPM's values in a plant historian. After several weeks, the plant personnel had determined the average weekly value of the measures for the baseline period and began using them as the starting baseline for measuring plant performance. After this baseline period, the plant personnel undertook several initiatives that they felt would positively impact their section's economic performance. Among other activities, they installed advance process controls, trained operators, and determined and disseminated best practices for their operations. The DPMs showed that the plant team's activities had improved the plant's economic performance by approximately $1.5 million annually. When this information was conveyed to the plant manager, he asked the plant's financial manager to confirm the results. Fortunately, the finance team in this plant was more open minded than most and worked diligently to detect the improvement. Several months later, the finance team confirmed that there had been a marked economic improvement in plant operations, on the same order as the operations team had reported, but they could not determine where the improvement came from. Again, it was fortunate for this plant that this general confirmation of improvement was sufficient to convince the plant management team that the operations team had accurately estimated the value that the performance improvements had contributed. Nevertheless, cases like these were demonstrating the need to somehow bring the DPMs and accounting systems together.

Converging DPMs with cost accounting systems would appear to be fairly straightforward. However, many of the same limitations in accounting systems that had led to the need for DPMs in the first place hampered efforts to bridge the gap between traditional accounting and DPMs. Much of the functionality of accounting systems has been defined by governmental reporting and auditing requirements, not by the operating requirements of the plant. Because of this, the accounting systems have not evolved to be well aligned with operational information requirements. A plant-centric view of a traditional cost accounting system is shown in figure 4.15. Although this is an overly simplified view, it demonstrates the gap between accounting and DPMs. To perform the plant's accounting functions, its accounting staff inputs the plant's total energy consumption, raw material consumption, and production volume over a given period (typically a month) into the accounting system along with other direct and indirect costs. The plant accounting system stores this information in its database, analyzes it according to predetermined rules, produces several reports, and provides summary information to the plant's enterprise resource planning (ERP) system for further analysis and reporting.

One of the reports generated by the accounting systems is the monthly variance report, which compares the expected standard cost per unit of product produced against the actual cost per unit of product produced. This type of variance report has been the most frequently used system for measuring manufacturing performance over the past century. However, this measurement approach has many basic flaws, which have been addressed extensively in numerous articles and books. But two of these flaws are particularly relevant to the effort to link these traditional accounting systems with the DPM information. First, the resolution or "granularity" of the data in traditional accounting systems typically stops at the total plant level. DPMs, on the other hand, provide cost and profit data at the plant-unit and plant-area levels, and the traditional accounting database provides no location to which information about these levels can be linked. The second disconnect involves timing. Cost accounting systems use a cost window of months, weeks, and in some advanced cases days. DPMs, by contrast, are calculated in real time to provide the timely information required to support operational decision-making. As plants implemented the early DPM systems and attempted to link them into the accounting systems, these two issues became significant barriers.

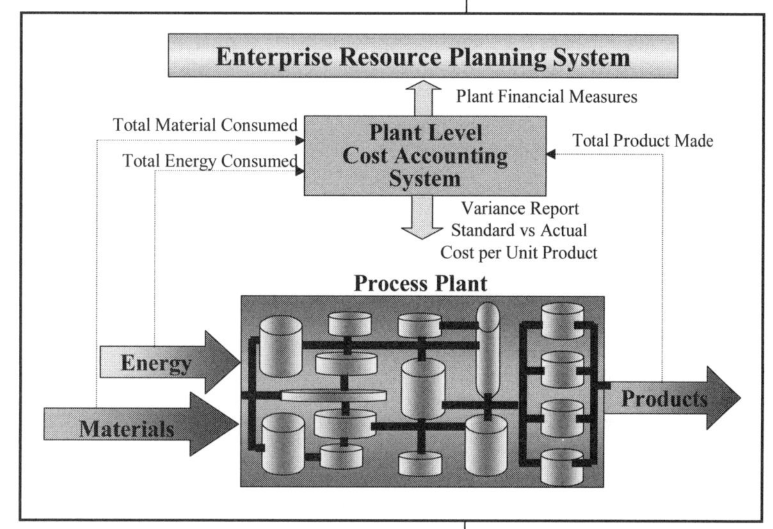

Figure 4.15 Current Plant Accounting Model

The advent of activity-based costing (ABC) systems offered plant personnel a ray of hope that DPMs could be linked to their plant's finance systems. (Activity-based accounting systems were introduced earlier.) As we have seen, this relatively new approach to manufacturing accounting was developed to try to compensate for some of the deficiencies in cost accounting for manufacturing. Although the concepts of ABC have been around for many years, ABC systems only really began to take shape and gain popularity in the late 1980s. ABC systems provide an alternative to the traditional general ledger-based accounting practices because they focus on the activities related to product manufacturing within an operation. ABC systems, for example, develop accounting views for each defined activity, which provides accounting data of much higher resolution or granularity. The manufacturing process equipment installed in a process plant defines the plant's primary manufacturing activities. Although most ABC systems developed in the past decade are targeted to discrete manufacturing operations, some—such as the Prism and Protean systems developed at Marcam and the Ross System iRenaissance package—were specifically designed with the characteristics of process manufacturing in mind.

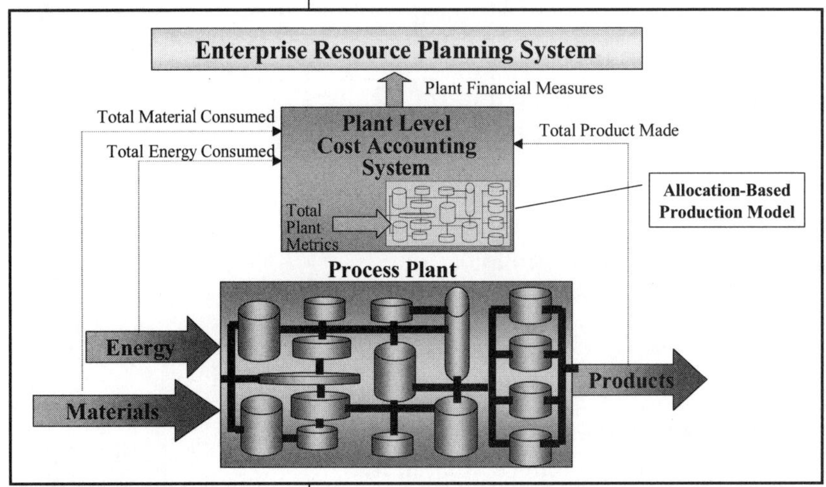

Figure 4.16 Production Model-based Accounting System

To model the activities in process plants these ABC systems have a production model (see figure 4.16) that provides a database view capable of matching the equipment layout of the process. This appears to be well aligned with the DPM approach since the basic activity level of a production model can be a process unit, which is also the basic level for DPMs in process plants. The problem, however, was that the process manufacturing plants in which the first ABC systems were implemented still used information on total plant resources (energy cost, raw material costs, and so on), so allocation algorithms had to be devised to allocate these total costs to the individual activities. For this reason, any improvement made to the operation of a single process unit had to be input into the ABC system as an improvement on plant totals and then allocated to all individual process units in accordance with the predetermined allocation algorithms. This cumbersome allocation approach made it very difficult to rationalize these ABC process systems with the DPMs.

Adding a production model to plant accounting systems was a major step forward, but most early implementers did not take full advantage of their new systems. Perhaps one of the reasons was that DPMs did not exist in the plants where the first ABC systems were installed. Without DPMs, the only reasonable approach to determining the costs and profits for the activities in a process plant was to allocate backward from the plant's total costs. With DPMs, the cost and profit of each activity (process unit) in a process manufacturing operation can be directly determined from DPM models, which use sensor-based data (figure 4.17). These calculated DPMs can be used to directly assign values into the ABC system's production model without having to rely on allocation algorithms. By directly assigning the values, the cost and profit information in the ABC system becomes much more representative of what is actually occurring in the manufacturing process.

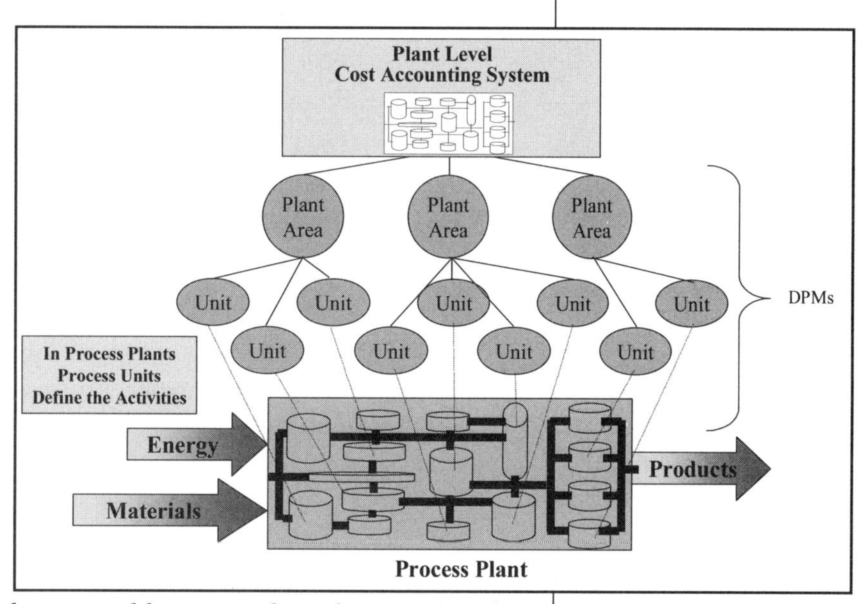

Figure 4.17 Performance-based Plant Accounting Model

The alignment of DPMs with process ABC systems provides a new level of operational information that can be tightly aligned with a plant's finance system. Unfortunately, ABC systems have not been implemented in manufacturing to any large extent thus far. There are several reasons for this. First, as the interest in and activities surrounding ABC systems began to grow in the late 1990s, the Y2K scare was commanding the attention of most IT and finance departments. Considerable cost and effort was expended on migrating to a standard ERP system that was Y2K compliant. The focus on new approaches, such as ABC, took a back seat. Second, ABC was often presented as an alternative to a traditional accounting system rather than a supplemental system that could complement ERP systems. Once manufacturing companies had invested all their time and effort into upgrading their ERP systems the last thing they wanted to hear was that they should replace them with a new type of system! Third, implementing ABC systems is much more complex than implementing traditional accounting systems. The individual activities have to be identified and measured, which adds a level of complication not required by traditional ERP systems. Largely because of these factors manufacturing companies have not moved to ABC systems as readily as was once expected.

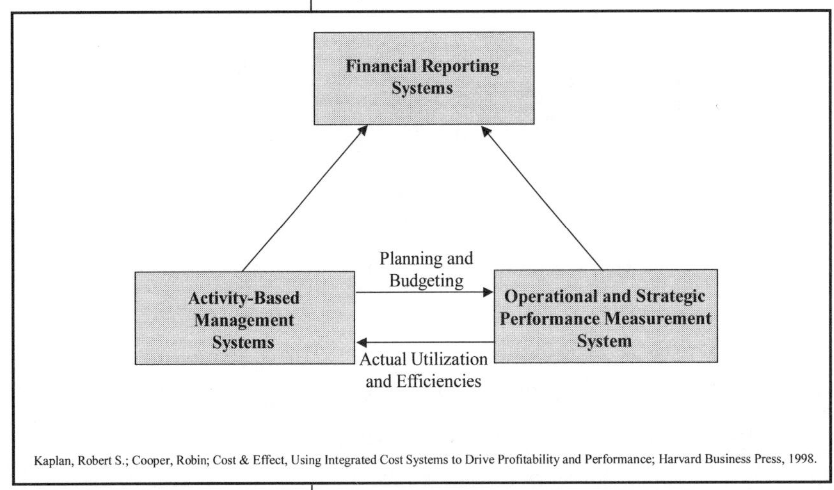

Figure 4.18 Model for Cost and Performance Measurement (Stage IV)

In more recent years, however, interest has grown significantly in ABC and its use as a decision-support and management toolkit (often referred to as "activity-based management" or ABM) as well as in new performance measurement systems. This may be indicative of a transition in cost accounting that will help to bring a much more holistic approach to the effective management of process manufacturing operations. In their groundbreaking book *Cost and Effect*, Robert S. Kaplan and Robin Cooper provide an excellent and compelling perspective on the evolution of cost accounting. They partition that evolution into four stages, evolving from the homegrown systems that dominated cost accounting throughout the past century through standard ERP-based systems all the way to what they predict will be the cost accounting approach of the future (they refer to it as "Stage IV"). Their assessment is that today's ERP-centric view of a single accounting system will not suffice. ERP-based accounting systems are absolutely necessary for fiduciary reporting, but they are not sufficient for supporting operational decision-making. Kaplan and Cooper do not view ABC/ABM systems as alternatives to ERP systems but as supplementary operational support systems. Their model of the Stage IV system (figure 4.18) projects the optimal structure for the next generation of performance-based accounting.

In the Stage IV model, ERP systems perform the functions of financial reporting that are necessary to meet fiduciary requirements, but they do not provide the information needed for operations decision support. They predict that to support a performance- based accounting approach along with the ERP systems, manufacturers will require not only ABM systems but an Operational and Strategic Performance Measurement system. DPMs provide the real-time, sensor-based operational and strategic performance measurements. These can be directly connected into the ABM system to assign costs and profits to the activities in the ABM database. They can also be used to provide a continuous learning opportunity for the operators, engineers, and maintenance personnel in the plant. Other performance measures that cannot be modeled from process sensors will be required to complete the Operational and Strategic Performance Measurement system. The ABM systems operate as entities independent of the ERP systems. They take input from the Operational and Strategic Performance Measurement system and use it to directly determine cost and profit at a subplant level. ABM systems are ideal for short- to medium-term production planning and scheduling and can provide performance targets for the Operational and Strategic Performance Measurement system.

In Kaplan and Cooper's Stage IV model the communication path they predicted extends from the ABM system and the Operational and Strategic Performance Measurement system up to the ERP system. This is exactly the opposite of the most common practice in manufacturing-related cost accounting today. The current common practice is to try to extract information from the ERP system to help drive the product-costing, ABC, and performance measurement (variance reports) systems. However, the Stage IV model makes more sense in that its premise is that cost, profit, and other critical operational information can be most accurately determined at the point at which it is realized – the manufacturing process. Thus, the accounting systems should be driven from this base information and not the other way around. If organizations implement a Cost and Performance Measurement System in the way stipulated by the Stage IV model their accounting information will not only be more accurate. It will also ensure that the organization is aligned with its strategy as long as the strategic performance measurements are used to help drive operations. This Stage IV model is clearly of greater scope than the DPM approach, but the two approaches are totally complementary. Kaplan and Cooper present the Stage IV model as an ideal that may become reality in the future. But with the advent of DPMs in process manufacturing it is already available today.

The combined DPM, ABM, and ERP approach appeals naturally to the common sense of managers striving to optimize the performance of their process plants. However, I should point out that very few companies have started to move in this direction. Implementing an effective ABM system is a very complex undertaking. Fortunately, in process plants the definition of the process's basic activities don't vary as much as they do in discrete manufacturing. This gives process manufacturing environments a level of stability once the ABM systems are implemented. Still, few companies have had the drive and patience to implement ABM systems, and the same is true for Operational and Strategic Performance Measurement systems and DPMs. The Vollmann Decomposition process presented in this book makes it possible to identify a company's strategic performance measures, which are the measures that should primarily be used for operations decision support. Identifying operational performance measures is a pure accounting task.

There is a degree of flexibility entailed in the performance measurement system that can further complicate the maintenance of the system. If market conditions or an organization's market strategy change, its strategic performance measures should change accordingly. Typically, this involves changing the priority of the measures, but it could involve more drastic alterations, such as adding new strategic performance measures. It is this complexity that seems to have retarded industry's adoption of these systems. What will overcome this barrier is the realization that the economic value these systems can contribute to organizations can more than outweigh the cost and effort of implementing and maintaining them. The case studies presented in this book provide preliminary evidence of this.

Process manufacturers have also been slow to adopt ABM systems because most commercially available ABC/ABM systems have been designed to meet the requirements of discrete manufacturing operations. Fortunately, as we have noted, a few systems, such as Marcam's Prism, Protean and Ross System iRenaissance systems, have recently emerged that are specifically designed for the requirements of process plants. These process-based systems may help to drive the acceptance of the ABM and ABC approach in process plants. They are also well aligned with DPMs. As more of these systems are installed and operational the economic value of a plant's activities will become clear to their finance departments as well as to the operations, engineering, and maintenance departments. This clarity and agreement on the economic value-added will enable the lean process organizations to focus their limited resources on the areas that offer the most potential positive impact on the plant and company. The result will be process manufacturing plants performing at a higher operational and economic level than was ever before possible.

NOTES

1. Berliner, Callie, and James A. Brimson, *Cost Management for Today's Advanced Manufacturing*. Boston: Harvard Business School Press, 1988, page 39.

2. Ness, Joseph A., and Thomas G. Cucuzza, "Tapping the Full Potential of ABC," *Harvard Business Review on Measuring Corporate Performance*. Boston: Harvard Business Press, 1998, page 54.

3. Berliner and Brimson, *Cost Management for Today's Advanced Manufacturing,* page 25.
4. Berliner and Brimson, *Cost Management for Today's Advanced Manufacturing,* page 93.
5. Henderson, Bruce A., and Jorge L. Larco, *Lean Transformation.* Richmond, Virginia: Oaklea Press, 1999, page 203.
6. Berliner and Brimson, *Cost Management for Today's Advanced Manufacturing,* page 34.
7. Berliner and Brimson, *Cost Management for Today's Advanced Manufacturing,* page 19.
8. Berliner and Brimson, *Cost Management for Today's Advanced Manufacturing,* page 26.
9. Berliner and Brimson, *Cost Management for Today's Advanced Manufacturing,* page 1.
10. Hiromoto, Toshiro, "Another Hidden Edge – Japanese Management Accounting." *Harvard Business Review,* July-August 1988, page 26.
11. Berliner and Brimson, *Cost Management for Today's Advanced Manufacturing,* page 2.
12. Berliner and Brimson, *Cost Management for Today's Advanced Manufacturing,* page 1.
13. Kaplan, Robert S., and David P. Norton, "The Balanced Scorecard – Measures that Drive Performance," *Harvard Business Review on Measuring Corporate Performance.* Boston: Harvard Business Press,1998, page 135.
14. Eccles, Robert G., "The Performance Measurement Manifesto," *Harvard Business Review on Measuring Corporate Performance.* Boston: Harvard Business Press, 1998, page 27.
15. Meyer, Christopher, "How the Right Measures Help Teams Excel," *Harvard Business Review on Measuring Corporate Performance.* Boston: Harvard Business Press. 1998, page 108.
16. Berliner and Brimson, *Cost Management for Today's Advanced Manufacturing,* page 15.
17. Berliner and Brimson, *Cost Management for Today's Advanced Manufacturing,* page 164.

18. Berliner and Brimson, *Cost Management for Today's Advanced Manufacturing,* page167.
19. Drucker, Peter F., "The Information Executives Truly Need," *Harvard Business Review on Measuring Corporate Performance.* Boston: Harvard Business Press, 1998, page 19.
20. Peters, Thomas J., and Robert H. Waterman, *In Search of Excellence: Lessons from Americas Best Run Companies.* New York: Harper & Row, 1979, page 7.
21. Drucker, Peter F., *Innovation and Entrepreneurship: Practices and Principles.* New York: Harper & Row, 1985, page 161.
22. Berliner and Brimson, *Cost Management for Today's Advanced Manufacturing,* page 6.
23. Berliner and Brimson, *Cost Management for Today's Advanced Manufacturing,* page 6.
24. Kaplan, Robert S., "One Cost System Isn't Enough." *Harvard Business Review,* January – February 1988, pagc 62.
25. Berliner and Brimson, *Cost Management for Today's Advanced Manufacturing,* page 167.
26. Drucker, "The Information Executives Truly Need," page 23.
27. Berliner and Brimson, *Cost Management for Today's Advanced Manufacturing,* page 165.
28. Zuboff, Shoshana, *In the Age of the Smart Machine.* New York: Basic Books, 1988, page 95.
29. Davila, Antonio, and Robert Simons, "How High Is Your Return on Management?", *Harvard Business Review on Measuring Corporate Performance.* Boston: Harvard Business Press, 1998, page 87.
30. Berliner and Brimson, Cost Management for Today's Advanced Manufacturing, page 163.

CHAPTER 5

The Convergence

The tools exist — here, now, today — to forge an entirely new paradigm for managing process plant performance. This paradigm is firmly rooted in the convergence of the historical trends the executives we interviewed identified: technology, quality, and accounting. Further, it can be forged into a dynamic management structure designed to drive performance in process manufacturing and enable the manufacturing operations, along with all other key organizational operations, to be utilized as key components of a company's strategy (figure 5.1).

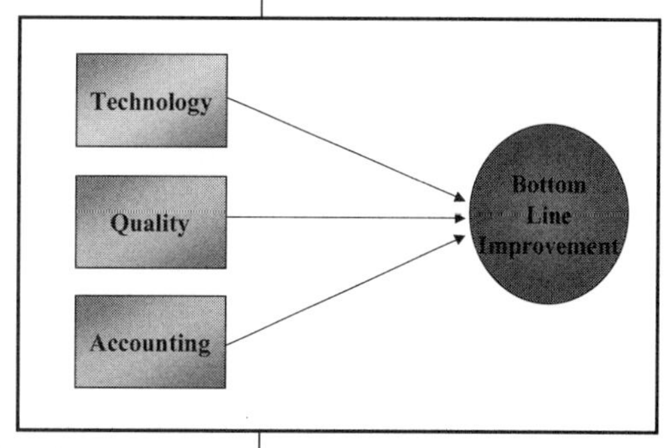

Figure 5.1 The Convergence: Survival in a Global Economy

Before describing this new management paradigm, it's appropriate to take a few moments to look at the relationship between the dynamic performance measures and the three historical trends we have discussed in this book. Dynamic performance measures provide the next step in the logical progression of *each* of these three trends and are, therefore, critical to their convergence. It will therefore be helpful to review each of these trends while we examine how dynamic performance measures provide the next logical step in the progression of each.

THE TECHNOLOGY TREND AND DYNAMIC PERFORMANCE MEASURES

Manufacturing managers have made significant capital expenditures in automation technology over the past thirty years, with little or nothing to show for it. Even with programs such as CIM, millions of dollars were spent with no discernible performance improvement. Many manufacturing managers have indicated that their plants are not running any better today then they were three decades ago, before the existence of computer-based automation systems. This is unfortunate and totally unacceptable. Manufacturing managers should be seeing real returns on all of their capital investments, and automation systems should be leading the pack.

The only way to get more payback out of automation investments is if the new technology provides increased functionality that improves the economic performance of the plant. The dynamic performance measures provide the information that will continually communicate the performance impact of any capital or operational investments. Automation suppliers have invested tremendous amounts of money to increase the functionality of their automation systems and software in order to provide manufacturers with improved capability to, in turn, drive the economic performance of their manufacturing operations. The expanded performance-enabling functionality embedded in the automation systems is seldom utilized because of the "replacement automation" and "project team" approaches that are almost universally used when implementing automation projects in process manufacturing environments. As pointed out earlier, the fact that there has been no measurement system in place indicates the economic impact from automation has actually shielded the lack of performance improvement from management. For years, manufacturing management has tended to approve capital budgets for automation projects on faith, assuming that they must be delivering reasonable returns. In the difficult competitive environment that globalization has spawned over the past decade, management's patience seems to be wearing thin. It can no longer afford to approve capital expenditures without a clear understanding of the economic value-added generated as a result.

When dynamic performance measures are implemented in these manufacturing operations, the "automation-with-no-measure-of-return" approach is bound to change. The dynamic performance measures provide the performance score cards that will continually report on whether the capital investments being made are providing economic returns. This fact should serve to drive the engineering and operations staff to get more performance from the automation investments by utilizing the advanced capabilities of these systems. Even the very limited data collected so far has demonstrated that the potential for improvement is enormous. But the improvement does not happen merely by acquiring a new, up-to-date automation system. As with any performance-enabling tools, the advanced features of the system must be used, and used effectively, to realize performance improvements. With the reductions in the size of plant engineering and operations staffs, the highly skilled personnel required to implement and manage the advanced automation features may not be available. This is causing automation suppliers to change the focus and scope of their market niches beyond just hardware and software companies by expanding their service offerings to fill the resource gap. Automation still offers one of the best opportunities for managing plant costs and production. Coupling advanced automation solutions with dynamic performance measures provides a proven way to continually increase the economic benefits of automation and improve the returns on the automation investments (figure 5.2).

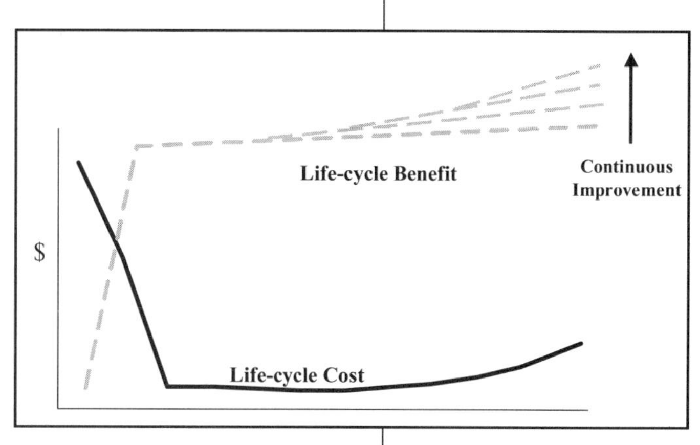

Figure 5.2 Improving Economic Performance

The effective implementation of dynamic performance measures *requires* the use of computer-based technology. As Robert Eccles has noted, "Information technology has played a critical role in making a performance measurement revolution possible."[1] For these measures to accurately reflect the current status of the manufacturing operation, the system that models them must be real time, and it must utilize the data about the manufacturing process provided directly by process sensors to generate the appropriate performance measure values at each level in the operation. The technology in real-time systems has become sophisticated enough to make the implementation of dynamic performance measure systems not only feasible but also actually quite simple, and very desirable. As the technology trend continues to evolve, organizations' ability to use automation technology to implement dynamic performance measures is driving the change from "technology for its own sake" to "technology for the bottom line."

THE QUALITY TREND AND DYNAMIC PERFORMANCE MEASURES

The quality movement has been growing very rapidly in the industrialized world for the past few decades. One of the key indicators of this is the increased use of SPC packages throughout manufacturing. It is unusual today to go into a plant that has not implemented SPC to some extent in its operation.

Many manufacturers in the process industry segments have tried to utilize the SPC packages that are popular in the discrete manufacturing segments. Although these implementations provide tidy and appealing SPC charts, the statistical techniques they employ do not translate directly to continuous process variables and they consequently produce suspect results. As Paul Badavas and Albert Epperly note, "For nondynamic processes such as parts manufacturing, the SPC discipline applies directly to the measurements and actuators. While for dynamic processes, such as distillation columns, SPC applies after Traditional Process Control (TPC) has driven the variables in question to steady state. The upper-level functions also profit from the statistical validity of data."[2]

This is not to say that the principles of SPC /SQC do not apply in process plants—they do. The point is that applying SPC/SQC in these environments requires careful and close investigation to achieve the desired continuous improvement.

Process plants have traditionally employed a much higher level of deterministic process control than that found in discrete factories. Because it has been prohibitive to provide deterministic control over discrete work cells, statistics were applied, and SPC/SQC evolved. This statistical approach was very successful at bringing the level of control in some discrete operations close to that of continuous process plants. But higher levels of control can still be realized. It is advisable to continually ask ourselves what aspects of the operation require the next level of focus for continuous improvement. In other words, what is the next level of quality indicators in these manufacturing operations?

If you walk into a process plant that has implemented a reasonable level of process control, and if you ask the operations and production managers what they need to continuously improve, their answer will often be quite simple: "plant performance." Although this may seem an obvious response, it is actually very significant: "Quality-related metrics have made the performance measurement revolution more real."[3] The key to success in today's marketplace is the performance of the entire plant, thus the plant quality teams need to find the next level of quality indicators (i.e., that are the indicators of the plant performance) at each point in the operation. These indicators are the dynamic performance measures for the operation. "Most executives have not yet realized, however, that such teams need new performance-measurement systems to fulfill their promise."[4] Until management recognizes this issue and takes the leadership necessary to drive the dynamic performance measures through their operations and use them to empower their manufacturing quality teams and plant personnel, the returns on the quality investments will certainly not be what they could be.

Christopher Meyers observes that "A well-designed financial control system can actually enhance rather than inhibit an organization's total quality management program."[5] The financially based dynamic performance measures provide the components of the well-designed financial control system that is required to achieve effective quality management in process plants. Although the financial metrics may be important in a quality management system, nonfinancial metrics are also very important. Many plant operations should have nonfinancial measures of performance, and these should be key components of the continuous improvement programs of quality teams and individuals in process environments.

A reasonable conclusion from this is that dynamic performance measures provide a new and effective approach for developing quality indicators in process operations and as such provide the key to the next logical step in the quality trend. Utilizing the dynamic performance measures as integral elements of a plant's total quality management program will make the program much more effective and facilitate the transition to quality for the bottom line.

THE ACCOUNTING TREND AND DYNAMIC PERFORMANCE MEASURES

Cost accounting systems have been the basis for measuring manufacturing performance for many years. Every month, manufacturing managers receive their variance reports informing them whether they did better or worse than expected in terms of the cost-per-unit of the products their operations makes. If the cost is consistently smaller than expected, they do well in the organization. If it gets larger, they do not do as well.

Today, much of the manufacturing world recognizes that the traditional cost-based performance measurement system is not effective or relevant for manufacturing operations. There seems to be a growing awareness that new measurement systems are required: "Activity-based costing represents both a different concept of the business process, especially for manufacturers, and a different way of measuring."[6] But even activity-based costing does not go far enough to meet the optimal manufacturing performance requirements.

It is generally accepted that the new measurement systems must take into consideration both the financial and nonfinancial aspects of an operation. Certainly, whatever the measures of performance are, they must match the plant's manufacturing strategy. People will perform to their measures. If the measures match the strategy, the strategy will be implemented; if the measures do not match the strategy, the strategy cannot possibly be implemented.

To ensure effectiveness, the performance measures must also be provided in a much more timely manner than those of traditional cost accounting systems. They must be provided in a time frame directly relative to the manufacture of the products. In other words, at least at the lowest levels of the manufacturing operation, they have to be provided in real time. The manufacturing personnel will not be able to improve their performance if they do not know how they are performing as they do their jobs. This real-time performance measurement approach is absolutely critical to their success.

Dynamic performance measures are real-time measurements of manufacturing performance that match the manufacturing strategy of the operation. The implementation of dynamic performance measures in a manufacturing operation results in either the separation of the manufacturing performance measurements from the cost accounting systems' performance measures or the fundamental restructuring of the way in which activity-based accounting is done. Either approach can provide excellent results.

Since traditional cost accounting has a set of performance measures built in that may not align with the dynamic performance measures, the most effective approach is probably the second: viewing dynamic performance measures as a fundamental restructuring of accounting for manufacturing. This new accounting approach can be referred to as "real-time, activity-based accounting" and can change accounting from an after-the-fact cost monitoring system to a performance-based, decision support tool. If the Vollmann Decomposition approach is used to develop the dynamic performance measures, this performance-based decision support tool will help drive consistency of purpose and performance throughout the operation. Real-time activity-based accounting provides a new accounting paradigm to promote accounting for the bottom line. Thus, dynamic performance measures provide the next logical step in the accounting trend that has been evolving over decades.

THE CONVERGENCE

When the same basic technology and/or approach provides the next logical step in the progression of multiple trends it is indicates that the trends are converging. This is exactly the situation we see with respect to dynamic performance measures and the technology, quality, and accounting trends. Traditionally, these three trends have been viewed and dealt with as though they were quite independent. The reality is that the trends and the drivers behind them have been very closely aligned, but since in most manufacturing organizations different departments have focused on each trend separately their alignment has often been overlooked.

A convergence of this type may appear to be innocuous on first inspection. But as Peter Drucker has pointed out, a "characteristic of knowledge-based innovations ... is that they are almost never based on one factor, but on a convergence of several kinds of knowledge, not all of the scientific or technological."[7] In today's market, knowledge-based innovations typically provide maximum impact on businesses, even though they may initially seem rather subtle. This is the case with dynamic performance measures and the convergence of the three trends: the implications of this convergence on manufacturing operations are very far reaching. They are leading to a fundamentally new paradigm for the management and operations of manufacturing facilities. For manufacturing operations to gain maximum benefit from the convergence they must implement the dynamic performance measurement structure and move to a new management structure, approach, and attitude.

A NEW MANAGEMENT PARADIGM

As we have seen, a new management paradigm is starting to evolve that takes advantage of the availability of dynamic performance measures. It is a naturally occurring management structure that results from the convergence of the three industrial trends discussed in this book (figure 5.3).

The objective of this new structure is fundamentally simple: to take advantage of all key manufacturing resources by combining them so as to maximize the performance of the entire operation. Fortuitously, the convergence of the three trends does exactly that.

This is fundamentally the same objective as the lean manufacturing approach so effectively implemented in discrete manufacturing operations. Many discrete manufacturing companies have revolutionized their manufacturing performance by effectively applying the six lean principles we discussed in chapter 3 (see figure 5.4).

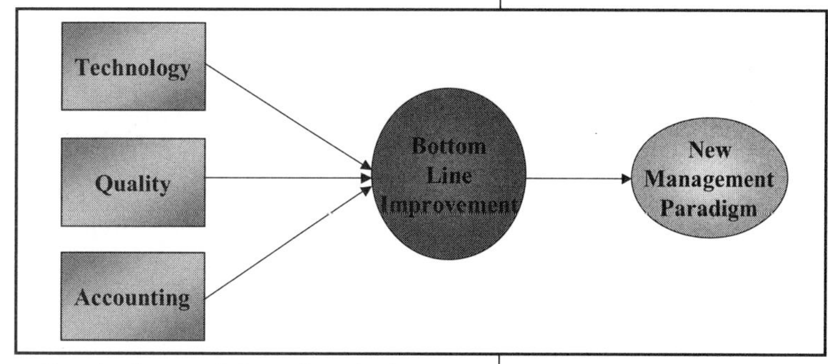

Figure 5.3 The Convergence

How to use the lean enterprise approach to achieve the same level of revolutionary performance in process manufacturing environments has been far from obvious. Though the characteristics of process manufacturing have made it very difficult to effectively apply all six lean principles in process operations, the availability of dynamic performance measures makes lean process manufacturing operations a possibility.

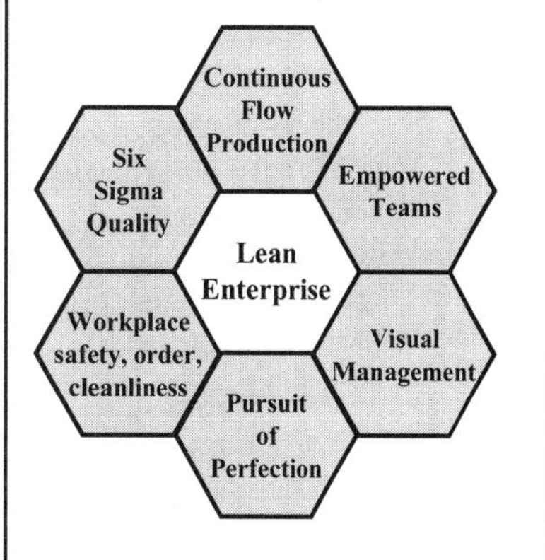

Figure 5.4 Lean Enterprise

The following is a short overview of how each of the lean principles may apply in process manufacturing environments and how dynamic performance measures actually enable the application of some of the principles.

1. *Continuous Flow Production* – As we pointed out earlier, most pure process manufacturing environments are continuous flow by nature and design. In fact, this is the one

lean principle that is a natural aspect of process manufacturing and thus that discrete manufacturing has had to emulate in order to become lean.

2. *Six Sigma Quality* – As we discussed in chapter 3, applying the principles of quality management in process plants has been an ongoing challenge. Using the dynamic performance measures as the quality indicators for plant performance helps to drive process manufacturing operations in the right direction. Clearly, it is difficult to think about process manufacturing in terms of defects per million, but the concepts of continuous improvement certainly apply. The statistic six sigma is not the key to this principle; a measurable way to drive the operation toward perfection is. Utilizing the dynamic performance measures as the quality indicators in a six sigma environment provides that drive for perfection.

3. *Empowered Teams* – The concept of empowered teams stems directly from the quality trend and TQM. To empower teams effectively, management must be able to measure their performance on an ongoing basis. The teams themselves also require their own ongoing performance measures. This has been the primary dilemma in empowering teams in manufacturing environments and ensuring their success. In process manufacturing plants that have implemented the dynamic performance measures, the basis for empowering the teams exists. The dynamic performance measures, established from the manufacturing strategy, are the measures to which the empowered teams should work. If the dynamic performance measures are decomposed out of the strategy and the empowered teams use them to measure their performance, management can feel comfortable that the fruits of the empowered teams' efforts will be very positive.

4. *Visual Management* – The dynamic performance measure performance dashboards offer a very effective methodology for providing visual management throughout manufacturing operations. The fact that the dynamic

performance measures are derived from actual process sensors means that the timeliness of the performance dashboard updates can be made to match the time constants of the jobs being performed. The dashboards of the front-line operators can update in seconds. This allows them to discern the performance impact of their actions as soon as they take the actions. As the dashboard displays for upper levels of management are developed, the time constants of the data update may be longer than that of the operators, but they should be right in line with the time constants of their activities. This provides the simple and clear visual management approach necessary to drive the lean enterprise concepts through process operations.

5. *Workplace Safety, Order, Cleanliness* – In process plants, the operator is often removed from the process and located in a control room. As a result, the concepts behind this principle may not translate as directly in process environments. As we noted in chapter 3, the control room environments are typically cleaner and more orderly than the manufacturing plant environments, but most control room environments could be significantly improved. Providing the information the operators need to be effective in a simple, clear, and orderly way on the operator workstations helps to implement this principle of lean production.

6. *Pursuit of Perfection* – The fundamental concept of "pursuit of perfection" implies that there is a set of identifiable and measurable objectives that essentially define what perfection is relative to current conditions. DPMs provide the basis for the definition of "perfection" in process manufacturing operations and help convert the pursuit of perfection from a nebulous concept to very specific and highly visible goals. Providing DPM dashboard displays across the entirety of a process plant helps to align all operations, engineering and maintenance personnel to the key performance objectives and helps to drive continuous improvement in the pursuit for perfection.

The Lean Production approach and concept have most commonly been associated with discrete manufacturing operations. Effective implementation of Lean principles can help to boost the performance of any manufacturing operation. Applying Lean Production techniques in process plants has been a daunting challenge, but with the advent of Dynamic Performance Measures the path has been cleared. The combination of DPMs and Lean Production provides a highly focused, performance-based process manufacturing environment in which every employee is enabled to perform to a common mission. Manufacturing may be a mature discipline but the combination of new technologies and management approaches can help to drive new and higher levels of performance even in the most traditional manufacturing operations.

NOTES

1. Eccles, Robert G., "The Performance Measurement Manifesto," *Harvard Business Review on Measuring Corporate Performance*. Boston: Harvard Business Press, 1998, page 32.
2. Badavas, Dr. Paul C., and Dr. Albert D. Epperly, "Statistical Process Control Integrated with Distributed Control Systems," 1988 NPRA Computer Conference, October 1988.
3. Eccles, "The Performance Measurement Manifesto," page 31.
4. Meyers, Christopher, "How the Right Measures Help Teams Excel," *Harvard Business Review on Measuring Corporate Performance*. Boston: Harvard Business Press, 1998, page 100.
5. Kaplan, Robert S., and David P. Norton, "The Balanced Scorecard – Measures that Drive Performance," *Harvard Business Review on Measuring Corporate Performance*. Boston: Harvard Business Press, 1998, page 136.

6. Drucker, Peter F., "The Information Executives Truly Need," *Harvard Business Review on Measuring Corporate Performance*. Boston: Harvard Business Press, 1998, page 4.

7. Drucker, Peter F., *Innovation and Entrepreneurship: Practices and Principles*. New York: Harper & Row, 1985, page 111.

CHAPTER 6

Case Studies

INTRODUCTION

Implementing a DPM system correctly requires a blend of strategic, accounting, and process engineering skills. It would be great if a nice fill-in-the-blank template existed to easily guide a practitioner in how to correctly identify the strategic performance measures and generate those measures' algorithms. Unfortunately, no such template exists.

The most critical aspect of a DPM implementation is identifying the correct measures. Even if the measures modeled and displayed on a DPM dashboard display are wrong, the plant's personnel will still tend to perform to those measures, which could drive the plant's performance in the wrong direction. Rigorously following the Vollmann decomposition process will help to ensure that the correct measures are identified, but the process still relies on the analyst's strategic, accounting, and engineering skills. We discussed the Vollmann decomposition process in chapter 4, but the best way to understand the process is to illustrate it with case studies of actual implementations. Each of the following case studies are based on real programs, and each illustrates important aspects of DPM implementations.

CASE STUDY 1: DYNAMIC PERFORMANCE MEASURE PROGRAM AT A MAJOR PHARMACEUTICAL COMPANY

The pharmaceutical company in this case study has become a leader in automation for automating its global bulk pharmaceutical processing facilities. Its management team has been driving a major corporate transition to make the company more competitive in the tough global marketplace while maintaining its strong emphasis on quality. Major components of the transition include refocusing the employees on those activities that provide the highest level of economic value-added (EVA^{TM} – the residual after-tax profit after deducting the full cost of capital. EVA is a trademark of Stearn Stewart). This shift has begun to influence all of the firm's employees, including its technology community. Company personnel have invested significant effort developing life-cycle economic models for the firm's automation investments and determining the economic value-added (EVA^{TM}) it is realizing from those investments. One of the major issues they face implementing management's objectives is how to measure the EVA^{TM} of the activities in the firm's manufacturing operations.

In late 1998, a series of discussions occurred between the pharmaceutical company's Global Automation Team and the Performance Measurement Team, which consisted of two business measurement consultants and two performance improvement engineers of The Foxboro Company. The subject of the discussions was how to measure the economic performance of the bulk manufacturing operations accurately and in real time. The impetus behind the implementation of the real-time performance measures was to demonstrate which activities in the plants add economic value and which do not. With this knowledge, the pharmaceutical company can make more intelligent decisions about where and how to invest its resources.

These discussions generated considerable interest among the pharmaceutical company's automation personnel, who have agreed to test the DPM concept to determine its applicability in their bulk pharmaceutical operations. One of the company's large bulk pharmaceutical plants has been selected for the implementation. It is a highly automated, best-in-class facility that performs quite well and is therefore determined to be an ideal site for this performance measurement program.

Overview of the Pharmaceutical Company's Facility

The selected facility manufactures bulk active pharmaceuticals for the global specialty pharmaceutical market, which are shipped to other facilities for final packaging. The plant's production technology is based on organic synthesis and product purification. The site has production buildings in which the pharmaceuticals are produced, a tank farm for supplying solvents to the production buildings, and an environmental controls section, which manages all aspects of the production waste streams. The environmental controls section is comprised of two areas for recovering solvents from the spent solvent slurry created from the manufacturing operations and two incinerators for disposing of the plant's liquid waste.

The facility's environmental controls area is a critical part of the site, and it was very clear to Foxboro's Performance Measurement Team that the pharmaceutical company was extremely concerned about its environmental stewardship. It applied considerable effort to minimizing the facility's environmental impact. In fact, the pharmaceutical company is so concerned about environmental issues that it has decided not to increase the number of incinerators on site, although this could constrain the plant's production capacity.

From the interview process, the Performance Measurement Team has learned that the pharmaceutical company's management has established several important goals for this production facility. The objectives are to

- continuously improve environmental stewardship
- continuously improve safety
- maximize productivity of the physical assets
- operate at minimum inventory levels

- ensure that manufacturing costs are lower than those of good generic manufacturers
- be the best value supplier in all aspects of customer service
- ensure that the site is the customer's first choice for new products
- aggressively understand and use compliance to drive long-term improvement

Understanding a site's goals is critical to any performance-improvement program. The plant had already worked very hard to provide measurements that are in line with the site's goals. All the personnel the Foxboro Performance Measurement Team encountered at the plant appeared to be quite aware of the measures and regularly referred to them in their daily reports. This strong measurement focus created an ideal environment for implementing, using, and accepting dynamic performance measures.

Performance Measurement Analysis at the Pharmaceutical Company

After several discussions, the plant personnel and the Performance Measurement Team decided that the best initial course of action would be for the Performance Measurement Team to visit the plant. The objectives of the visit were (1) to present and discuss the dynamic performance measurement (DPM) process with the plant's management team, (2) to select a plant section, and (3) to conduct a DPM analysis on the selected plant section. The visit at the facility was scheduled over a two-day period in March 1999. The primary representatives of the pharmaceutical company who were involved in the analysis were the leader of the facility's Process Control Group and an engineering consultant from its Engineering Tech Center.

The Performance Measurement Team's first step once on site was to do a presentation for the plant's management team on DPM and the analysis process. The presentation was critical because it created alignment between the Performance Measurement Team and the plant's management team and set expectations with respect to the analysis process and overall DPM process.

As a result of the presentation some members of the team raised a concern that in an FDA-regulated plant such as this plant, the operators may not have the necessary freedom to drive the improvements once the performance measurements are in place. Once the firm and the FDA agreed on good manufacturing practices (GMPs) for the facility, they would have to be followed to the letter. The concern was that the FDA would have to certify any improvement before it was implemented. The joint team acknowledged this limitation and agreed that the FDA regulations could significantly limit the plant's improvement potential. On the other hand, the team determined that some improvement might be possible even within the definition of the GMPs, and so the program was worth a try.

The Performance Measurement Team then began to undertake a structured series of interviews with plant personnel. These interviews are the primary on-site activities underlying the Vollmann decomposition analysis. They were conducted in such a way that the person being interviewed did not need to understand the intricacies of the decomposition process. They only needed to respond to a few key questions. The interviews were designed to identify the strategic performance measures that reflect the plant management team's strategic goals and objectives and to decompose those measures down through the various plant sections to the process units.

The leader of the Process Control Group was selected to represent the plant on the Performance Measurement Team. His participation helped the team to synthesize the information it received from other company personnel and to provide any additional information it needed.

The Vollmann decomposition analysis process provides a top-down approach to identifying and decomposing the facility's strategic performance measures. Therefore, the first interview was conducted with the site's general manager. During this interview, the plant's key performance measures were clearly identified. The primary measure was plant production capacity, which is the measure of potential plant production in a given period of time. This measure was a crucial piece of information because the plant is but one of a number of the pharmaceutical company's plants that were designed to manufacture bulk pharmaceuticals. When its R&D group develops new products management determines which of its plants will manufacture them, based on the plant's manufacturing capability and its available production capacity. Thus, the economic profile of any plant is directly driven by its production capacity.

As the interview with the general manager progressed, the plant team's concern about the environmental impact of the plant on the local community deepened. Local environmental concerns are very strong throughout the areas in which the pharmaceutical company operates, so some environmental measures were determined to be essential considerations for this plant. Several times during the interview, the GM also pointed out that the plant's environmental controls section was operating at capacity, so they would not add more incinerators to expand the production capacity. The Performance Measurement Team initially interpreted these comments to mean that the plant's environmental control and incineration sections were a major constraint on the plant's primary performance measure: production capacity.

The plant team's other interviews were conducted with several plant operations managers as well as with engineering personnel. The objective of these interviews was to understand the plant's layout and structure, to ascertain the organizational decomposition of the plant's operations teams, and to gather information for use in the DPM decomposition analysis. The team also toured various plant areas throughout the day.

The interviews demonstrated that the plant's management team had a solid understanding of the concepts of DPM. The metric program that was already in place at the plant, which had already been decomposed to the level of the plant's processing sections, provided a good basis for the DPM program. Each manager produced summary reports to determine the daily or weekly improvement on each identified metric.

As the interviews progressed, several members of the plant's team suggested that the Performance Measurement Team should focus the test DPM analysis and implementation on the plant's Environmental Controls area. This suggestion seemed to be in line with the issues expressed in the interview with the site's general manager. The solvent recovery and incineration sections were suggested and selected for three important reasons:

1. The management team was keenly aware of the plant's impact on the local environment. Since a significant portion of the plant's environmental impact derives from the recovery and incineration operations, these operations are critical to plant management. The local office of the Environmental Protection Agency (EPA) requires that emission measurements be made continually to gauge the environmental impact of the plant. With this in mind, the pharmaceutical company has decided to limit the plant's incineration capacity to the two existing incinerators. If the incineration capacity is exceeded, they will transport the waste off site for incineration, at considerable expense.

2. With the plant's incineration capacity limited to the two existing incinerators, its recovery and incineration capacity is becoming a major production bottleneck.

3. The alternative areas, namely, the pharmaceutical buildings, strictly adhere to FDA cGMPs (current good manufacturing practices), and plant personnel feel they may not have the necessary freedom to achieve improvements in these areas of the plant.

The joint team agreed to focus the remainder of the analysis on the facility's Environmental Controls area, which, as noted, consists of the solvent recovery and incinerator sections. This focus was not intended to imply that these were the only areas of this facility that would benefit from real-time performance measures. Nor does it mean that these plant sections would provide the highest economic returns. Rather, this focus was intended to establish an initial area of concentration that meets the requirements of both the pharmaceutical company and the Performance Measurement Team.

Process Overview and DPM Analysis

The facility's solvent recovery section and incinerator section are tightly linked from a processing and from a DPM perspective. A spent solvent mixture containing the various solvents used in the production process as well as some residual waste from the dry ingredients is output from the production buildings. The spent solvent slurry is transferred from the manufacturing buildings, through the tank farm, to one of the two solvent recovery sections. The slurry is processed through distillation columns and stills in the solvent recovery section in order to reclaim as much as possible of the original solvents in a pure enough form to be reused in the production processes.

The primary objective of the Environmental Controls section is to *process as much solvent slurry as possible so as to reduce the processing cost and increase the overall production capacity of the plant,* within the plant's environmental constraints. The Environmental Controls section also produces very high quality solvents for reuse in the processing steps of the production buildings.

There are three basic elements of the solvent recovery section's mission: (1) to recover as much as possible of the different solvents used in the plant, (2) to meet regulatory requirements, and (3) to manage the supply of solvents to the operations. Increasing the amount of recovered solvent reduces the amount that needs to be processed through the incinerators. This has two effects on the facility. The first is an increase in the availability of solvent for operations. The second is a reduction in the amount of waste generated, which is critically important to Lilly because of its commitments to the local community's environment and the EPA.

Each of the solvent recovery sections consists of a number of distillation columns and stills, which are used to recover the solvents from the solvent slurry stream (see figure 6.1). The distillation columns must be "tuned" to the specific characteristics of the solvent slurry that is being treated at the time. Since the manufacturing operation involves multiple products and campaigns, the operators may need to adjust the "tuning" of each column frequently. The start-up of the columns for each slurry stream is a manual operation that typically takes from three to five hours, depending on the operator. Until a column is up to full production and efficiency, the column's output goes back into solvent slurry storage to be retreated.

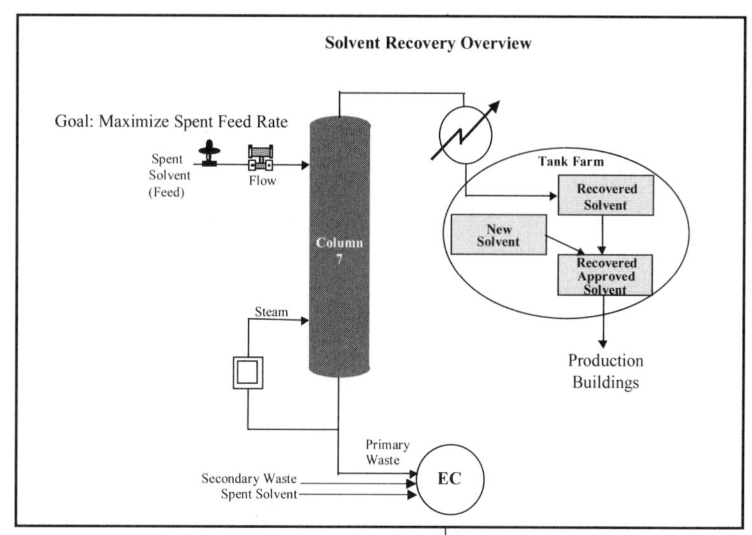

Figure 6.1 Solvent Recovery Overview

The selected solvent slurry is introduced into the distillation column from a spent solvent in the tank farm. The composition of the feed is determined through lab analysis. The integrity of the composition of the feed stream is extremely important to the recovery process. Some solvent slurry feed streams need to be processed many times to achieve a solvent of sufficient purity to be reused in the operations. Solvents are recovered through distilla tion and transferred to a recovered solvent tank in the solvent tank farm. Lab analysis is performed on the recovered solvent to determine if it is at the quality level that has been approved for use in the manufacturing operations. If approved, it is then transferred to a recovered/approved tank and can be reintroduced into the manufacturing buildings as appropriate. Solvent that does not meet quality standards is reintroduced into the recovery process.

There are two incinerators at the plant, a John Zink incinerator and a Thermall incinerator. Both work on the same basic principles (see figure 6.2). Two waste streams are introduced into each incinerator, a primary waste stream and a secondary waste stream. The primary waste stream contains the nonrecovered solvent output from the recovery process. It consists of solvents and residual materials from the manufacturing operations and is highly flammable. The primary waste is introduced as fuel into each

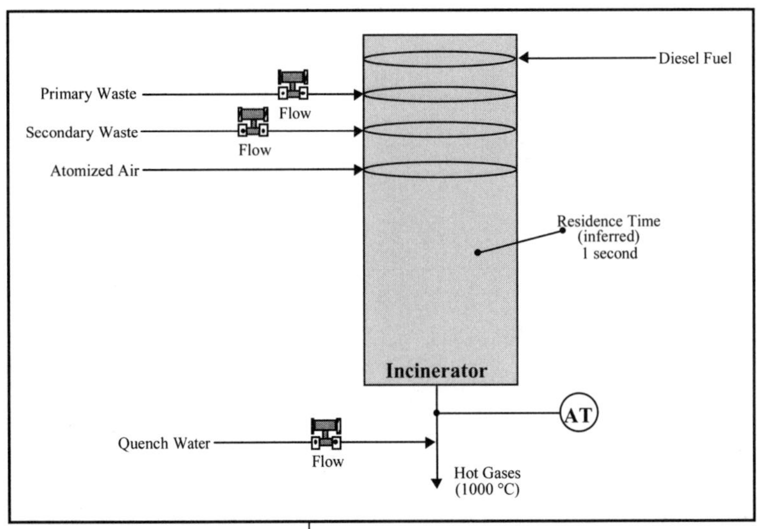

Figure 6.2 Incinerator Overview

incinerator and is used to incinerate the secondary waste stream. The secondary waste stream is comprised of production waste, such as the water-based residue generated when the process equipment is cleaned. The secondary waste has a high water content that requires a significant amount of heat to incinerate. The temperatures in the incinerators are maintained at just over 1,000°C. A residence time of one second at this temperature is required to incinerate the secondary waste. Atomized air is also introduced into the incinerators to stimulate combustion. Diesel fuel is used to help start up the incinerators and to compensate for any shortcomings in the BTU content of the primary waste.

Analytical instruments are used to determine the emission levels in the flue gasses (HC, CO, HCL) and to ensure that they comply with EPA limits. Quench water is used to pull some of the impurities out of the flue gasses and accelerate cooling. When the incinerators reach certain environmental limits (HC, CO, HCL), they will trip, which causes them to automatically shut down. Once an incinerator shutdown occurs it takes the operations staff some time to restart the incinerator and return it to the operational limits. The primary operational objective of the incinerators is to be operated at the highest rate of incineration possible with the lowest possible environmental emissions.

The efficiency of the combined solvent recovery and incineration processes is the key to the site's performance effectiveness. The Performance Measurement Team suggested implementing and managing appropriate dynamic performance measures for each of the solvent recovery and incineration sections. Although there are a number of economic performance metrics that could provide indicators for these important areas of the operation (e.g., power costs and recovered solvent value), optimizing these variables independently would not address the plant's chief concerns. In fact, optimizing some of these variables could lead to operator and engineering actions that actually decrease the economic value of the overall facility. For example, it was suggested that one appropriate measure might be calculated as the percentage recovery divided by the cost of recovery. The objective would be to increase this metric through a combination of actions: increasing the numerator and decreasing the denominator. The best way to increase this metric may actually be to reduce the denominator (power cost)—which is not a major economic consideration to the plant—relative to production value. Optimizing this metric could actually reduce the production capacity and production value of the plant, which would have dire economic consequences for the overall facility.

The next aspect of the DPM analysis was to develop an operational decomposition of the site to determine which control rooms and operating areas would need to have DPM dashboard displays presented to the operators. The actual DPM construction is accomplished for each operational unit in a plant, but the dashboard displays are constructed for an operator's functional domain of responsibility. It is therefore important to map the operational measures to the operator displays. There is an operating super-intendent for the overall Environmental Controls section, but there are two control rooms under that superintendent. One of the control rooms hosts the operating team for both recovery operations, and the other control room is for both incinerators. This operational structure led the team to the conclusion that two sets of DPM dashboards would need to be developed one for the solvent recovery operators and the other for the incinerator operators. Once this information was collected the Performance Measurement Team left the site to compile the DPM analysis report.

DPM Analysis Report

The Performance Measurement Team developed the DPM analysis report and sent it to the plant management team for review within a month of the first site visit. After reviewing the initial report, the plant team members explained that they understood the logic of the Performance Measurement Team's perspective on production capacity. The plant management team explained that the value of the products made at the plant was so great that they would never allow the Environmental Controls area to become the limiting factor on production capacity. Once this was understood, it became clear that the value of increasing the capacity of the Environmental Controls section was the cost savings that would result from not having to send out the spent solvent for external processing vis-à-vis the value of the increased production capacity.

The type of interaction that took place between the Performance Measurement Team and the site management team after the initial review of the DPM analysis report is essential, and changes at this point in the process are not unusual. The Performance Measurement Team has found in its experience that most manufacturing operations and engineering groups are not comfortable dealing with econometrics. So making some econometric assumptions and presenting them back to the plant team is a useful way to get the team to understand what information is required and how important it is to select the correct values. Once the initial proposal was sent to the plant, the plant team's approach to the analysis changed. The plant team and the Performance Measurement team appeared to begin working more effectively together at this point because they had a common understanding of what the program's intended scope was.

The DPM analysis report was corrected based on a second round of discussions that occurred after the initial report was submitted. The plant team analyzed the second report and after a few weeks agreed with the proposal to implement the DPMs in the Environmental Control section of the plant. The Performance Measurement Team then paid a second trip to the site to build the DPM models, install them on the operating automation system, and historicize them using the historical database running in the automation system.

After the Performance Measurement Team made the requested adjustments, the primary objective of the DPM implementation was to enable the plant operators in the plant's solvent recovery and incineration sections to increase the capacity of the spent solvent processing. The econometric calculations for the solvent recovery and incineration sections were based on (1) the reduction in cost to the plant because it no longer had to pay for off-site incineration of spent solvent and (2) the increased value of the incremental recovered solvents.

The capacity of the spent solvent processing is determined by the serial relationship between the solvent recovery operation and the incineration operation. The first part of this relationship is the solvent recovery section's ability to recover more solvent from the spent solvent streams. The recovery of this additional solvent means more solvents are available to the plant for the manufacturing operation and less waste has to be incinerated. This metric is calculated as a function of the increase in recovered solvent.

The cost of the plant to process the waste solvent off site was determined to be a constant value *v1* per liter. The model is as follows:

Recovered Solvent – Baseline Recovered Solvent * v1

The second part of the serial relationship was the increased capacity of the incinerators to handle waste. For each liter of increased incineration capacity, one less liter had to be sent off site for processing. This metric is calculated as follows:

(Incinerator Throughput – Baseline Throughput) * v1

The total economic value is the sum of the reduction in waste through recovery, the value of the incremental recovered solvent, and the value of the incremental waste processed by the incinerators. Hence,

Value = Value of the Increase in Environmental Controls Throughput
+ Value of the Increase in Returned Products from Solvent Recovery

Two additional measures proved to be valuable: incinerator uptime (averaged over a twenty-four-hour shift day starting at 8:00 am) and amount of waste product from columns. The incinerators are a significant bottleneck in the process. If they stop processing solvents then the plant's ability to dispose of waste is reduced and the costs are significantly increased. Such an interruption in processing can be caused by either process and mechanical trips or by interlocks initiated by environmental monitoring instrumentation on the flues. Finally, reducing the total waste by recovering solvent would provide greater throughput of solvents in the Environmental Controls section, assuming the incinerators are being used on a constant basis. These metrics are not really econometrics, but they provide valuable information that helps the operators manage their prime econometrics.

Models of the DPMs and supporting operating measures were developed to operate in the automation system installed at the pharmaceutical company's site. Automation systems are natural platforms for implementing DPMs since they have direct access to the process sensors and have software that enables algorithmic models to be developed that operate in real time in the automation systems' control packages.

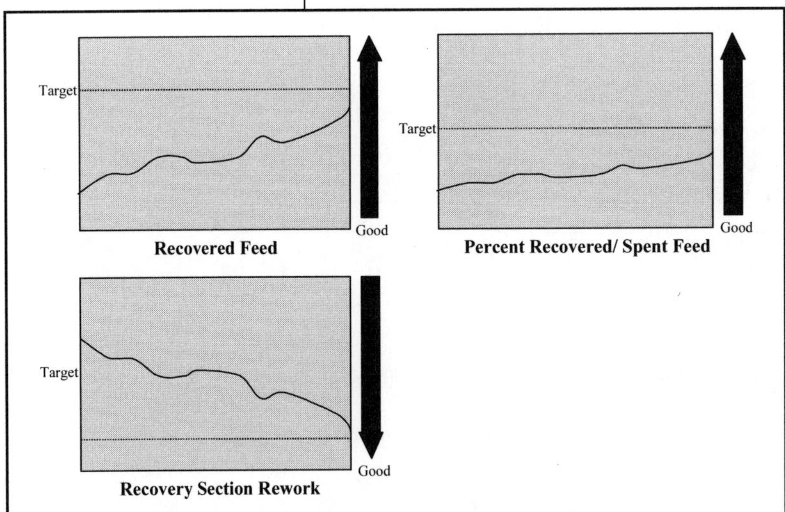

Figure 6.3 Solvent Recovery Operator Dashboard

Since the solvent recovery and incinerator sections of the operation are run by different operating teams, two DPM dashboard displays were developed for the Environmental Controls area at the site. One display was developed for the solvent recovery operators and one for the incinerator operators. Supporting metrics were developed for each section to provide additional information that would help the operations groups maximize the value of the primary metric. In each recovery section, the primary metric was the increase in recovered feed of the entire facility. Secondary metrics were also provided for increased percentage of recovered solvent to spent feed and reduced rework volume.

The primary metric of the incineration section's dashboard display was increased incineration throughput. This throughput is constrained by the level of contaminants released into the environment—an absolutely critical constraint on the operation of the plant. The operation will shut down if the limits are exceeded and the cost of a shutdown is excessive. Therefore, a second supportive metric is incineration uptime (see figure 6.4). Clearly, more time available for incineration will result in higher levels of throughput. A single incinerator trip could lead to an extended shutdown period since it takes time and effort to get the incinerators back on line and operating efficiently.

The Performance Measurement Team implemented the DPMs for the plant's Environmental Controls area during a second visit to the site in September 1999. They also implemented the historian data collections for each of the DPMs and calculated the monthly averages for each DPM into the historian. These monthly averages were used to establish the baseline values for the initial performance. The value of any performance improvement activities could be measured by their incremental deviation from the baseline operating value. The operator's dashboard displays were not implemented at this point so a reasonable baseline could be established for the metrics. The dashboards are actually performance-improvement support tools, and it would be impossible to derive valid baseline performance while the dashboards are operating.

The baseline values were established during a six-week period that ended in the middle of October. Once the baselines were determined, the plant team decided that the first performance improvement activity they would undertake would be to apply advanced process controls to one of the distilla-

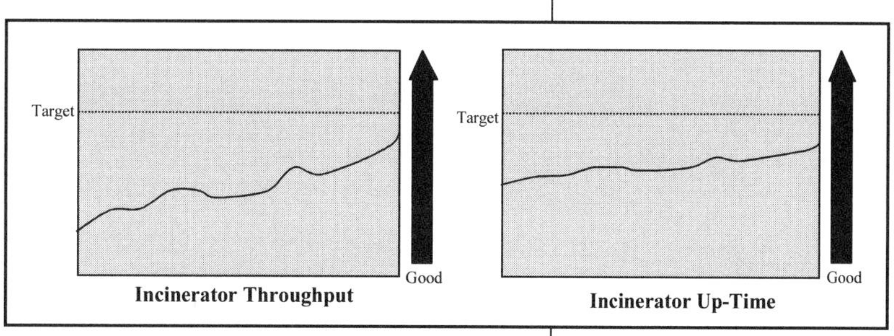

Figure 6.4 Incineration Operator Dashboard

tion columns in the solvent recovery section. The advanced controls were implemented, and the DPMs indicated that the annual value of this activity would add a sustainable $400,000 per year going forward. Without the DPMs it would have been difficult to determine the value of the advanced controls.

The Performance Measurement Team made a third site visit in January 2000. During this visit it implemented the DPM operator dashboard displays and trained the operator teams in how to use them. The plant team then undertook a series of other improvement activities involving new operating practices and operator training. Once again, it would have been very difficult to determine the value of these activities to the pharmaceutical company before the DPMs were implemented. With the DPMs operating, the plant team was able to obtain immediate feedback regarding whether an activity was providing incremental value. By using this information they were able to direct the operations staff to focus on those activities that were providing the most value. The net result was an incremental sustainable EVA™ to the plant operation of approximately $1 million per year.

The plant's performance improvement program is ongoing. The initial results of about $1.5 million of annual economic benefits to the pharmaceutical company have been very positive, and further improvements are expected to provide even better results.

CASE STUDY 2: DYNAMIC PERFORMANCE MEASURES AND CULTURE AT DYNEGY MIDSTREAM SERVICES

The bulk of the initial research that led to the development of the DPM approach and methodology was directed toward automation and information technologies, accounting systems, and quality improvement. As DPM and performance improvement programs were executed at various manufacturing plants, it began to become evident that it was necessary to determine and implement the correct DPMs in an operation and build the DPM dashboards for real-time operations feedback. However, it also became apparent that there were not enough components for an effective continuous performance improvement process. Once the operations management teams agreed on the appropriate DPMs and the DPMs were implemented, more often than not they were not utilized as effectively as anticipated. The success or failure of the DPM approach was directly tied to the culture of the manufacturing organizations. Generally, DPMs were most effective in team-oriented, empowered, accountability-based cultures.

The management team at Dynegy Midstream Services clearly understood these cultural issues and drove a company wide transformation that was designed to enable continuous economic performance. Dynegy Midstream Services is a leading-edge natural gas liquids subsidiary of Dynegy Inc., a $29 billion global provider of energy and communication solutions. The management of Dynegy Midstream Services strongly believes that its employees are one of the primary critical factors for Dynegy's success. This belief is so strong that Dynegy's corporate value has been defined as "We Believe in People." Dynegy's success in its key markets is evidence that its management's belief in employees is very effective. Through this focus, Dynegy Midstream Services' management team clearly views automation and information technology as the keys to enabling its people to improve Dynegy's performance: As Tom Fiske notes, "Dynegy Midstream Services is on the leading edge when it comes to integrating technology, information and corporate culture."[1]

The market for natural gas liquids had been undergoing tremendous change in the years before the launch of Dynegy's DPM program in 1999. Previously, natural gas liquids had been an integrated component of the business of the major oil companies. When many of these companies decided to exit this market, a number of new competitors emerged, among them Dynegy. Rather than the traditional service focus of the oil companies, these new businesses have a very strong focus on profitability. This led to rapid consolidation and the rationalization of asset portfolios. Basically, the natural gas liquids market transitioned from a side business of giant companies into the mainstream business of smaller, very aggressive companies. Survival in this segment requires operational excellence and unsurpassed economic performance.

Stephen Furbacher, Dynegy Midstream Services' president and COO, together with his team became interested in the concept of real-time performance measurement systems because they were so well aligned with the direction in which Furbacher was driving the business. Dynegy had a clear, well-communicated strategy based on the following goals:

- being the low-cost provider in its market segment,
- optimizing its asset base to extract value across the liquids value chain, and
- enabling an entrepreneurial culture.

Dynegy presented an extremely interesting case in the utilization of DPMs primarily because of its third strategic objective, "enabling an entrepreneurial culture." While conducting Vollmann decomposition analyses in several manufacturing operations, The Foxboro Company's Performance Measurement Teams had encountered many corporate strategic objectives. However, they had seldom encountered one as strong as Dynegy's cultural objective. On several occasions, business executives of other companies' manufacturing operations had told Performance Measurement Teams that they felt their internal culture might hinder the success of a proposed DPM program. In contrast, when Dynegy Midstream Services' DPM program was initiated the firm had already recognized the importance of corporate culture to its strategic success.

Furbacher and his team had started a very aggressive program to drive the company's business to a new operational and cultural model. His management team agreed that small incremental changes would not be sufficient; a radical change in business paradigm was required. The management team decided to move from the firm's traditional hierarchical, command-and-control management structure to a flat, team-based, empowered organization. As Fiske comments, "The company's restructuring effort included a transition from a six-strata hierarchical management structure to a much flatter two-layer self-managed organization. In what they describe as a benevolent dictatorship, the lower 'worker management' layer is free to determine how to achieve the goals and vision of the organization, yet it is still the top layer that defines these goals and vision."[2]

The functional organization transitioned to a performance-oriented organization. Performance-based compensation plans were implemented throughout the operation: "The management provides, but also fully endorses, encourages, and empowers workers to use information tools to fulfill their goals. The information tools not only provide workers with valuable information to perform their jobs, but also help to form a culture of collaboration and cooperation among individuals and groups."[3] It is easy to see why a metric-based DPM approach that uses automation and information technologies to deliver empowering, real-time information to employees was in tight alignment with the Dynegy management directives.

One principle that was very important to the Dynegy management team in moving to the new "entrepreneurial culture" was that this new culture required their operations to adopt a very different management approach. Dynegy's management team did not put the onus of driving the entrepreneurial culture on its workers; they put it on themselves. Team members determined that they had to invest in new skills for their employees and develop the support systems that make possible effective utilization of the new skill set. They created accountability matrices and allowed their employees to manage the business within their domains of accountability. They also encouraged employees to identify and focus on the "big rocks" that would generate the most value for the corporation. Furthermore, they tied individual incentives to the metrics of each individual's domain of authority and responsibility, and they encouraged a "multi-skill" human development program to broaden the perspectives and skills of their employees. The management team recognized that it needed effective real-time information so employees could continually discern their performance and improvement opportunities as they performed their day-to-day activities. Finally, the team provided the front-line employees with the skills and tools to manage their segment of Dynegy's business and then entrusted them to do the job. As Fiske notes, "Top management at Dynegy fully endorses a culture that not only empowers but also expects employees to respond to situations using the new information to increase productivity."[4]

Dynegy worked with Foxboro to implement DPMs in critical areas of its operations. As with every DPM analysis, the Performance Measurement Team conducted a series of top-down interviews: "Determining DPMs top down guarantees that they meet the strategic objectives of the company."[5] The results were very positive. Dynegy found that the availability of real-time information and its employees' flawless execution were the critical success factors. "DPMs effect nearly everyone at Dynegy Midstream Services directly or indirectly," Fiske comments. "DPMs provide Dynegy with a condensed set of accurate, reliable, contextual and visual targets with associated actionable items that affect performance and profitability targets. Most importantly, DPMs provide critical and timely information to maintenance workers, operators, managers and top executives that allow them to make more knowledgeable and faster decisions."[6] Dynegy has effectively utilized technology to enable and reinforce its cultural change. It runs its plants as businesses with team-based structures that extend right down to front-line plant operations. All personnel in the Dynegy Midstream Services plants are aligned with the business plan and are totally focused on executing it. DPMs help to facilitate this alignment.

Dynegy Midstream Services has made significant progress in moving its culture and operational performance forward, and the DPM approach has been an important ingredient in their success. In the first year of this program, they reduced operating expenses and G&A by $58 million, reduced maintenance capital spending by $6-8 million, and, on average, awarded $4,800 in incentive-plan bonus payments to their hourly operations and maintenance personnel. Since that first year, the savings have continued to grow, and Dynegy's safety record has improved dramatically. The result has been continuing annual incentive bonus payments to hourly employees.

Dynegy's management team is quick to point out that these improvements are not the result of any single initiative. Rather, they come from a comprehensive program that combines technology and people in a new team-oriented culture in which the teams perform many of the functions traditionally reserved for management, such as doing performance reviews and setting compensation levels. The DPMs are a part of the total program that ensures that the teams are receiving reliable, actionable information quickly enough to enable them to accomplish their activities in the most effective and efficient way. As Fiske confirms, "Dynegy Midstream Services is experiencing significant savings from improved operational performance by actively promoting a corporate culture that empowers workers to use DPMs to perform tasks that directly influence profitability."[7]

As Dynegy has progressed through this process it has learned several very important lessons that can help others embarking on a similar performance-improvement program. First, the cultural and people aspects of success were bigger than they had initially believed. Getting the entire organization to align itself with corporate strategy, vision, and value is critical to success. Real-time performance information, such as DPMs, enables the empowered teams to execute the strategy and reach established performance goals. This is a good example of how technology should be deployed as an enabler, not as an end unto itself. "Using DPMs, Dynegy has successfully coupled technology and information to people, which is a crucial element in extracting maximum value from their investments. By actively pursuing an atmosphere that integrates technology, information, and culture, Dynegy is experiencing significant operational performance enhancements."[8]

LESSONS LEARNED

Several very important lessons can be drawn from the case studies we have discussed as well as from the many other DPM implementations undertaken in other process manufacturing facilities. It may be useful to summarize some of these issues and discuss their implications.

One interesting phenomenon occurred at more than one site in which DPMs were implemented, including one of the sites covered in our case studies. This was that even though the users clearly identified sustainable economic value and the operational use of the DPMs, after some time the plants' operations teams turned off the dashboard displays. This behavior was unexpected, and evaluations were conducted to identify the reasons. In more than one case, the plant team indicated that the economic value improvement, which was over $1 million, was too small for them to spend the time and effort required to manage it. There were more important issues in the plants, namely, major capital projects in other parts of the plants that required both the plant engineering and operation teams' attention. Although this explanation was determined to be legitimate, the plants' personnel agreed that they still needed to man the operating station for the plant sections that had DPMs. They also agreed that reviewing the DPM displays did not consume significant resources.

After further discussion another important cultural issue was identified. Plant personnel related that management tended to reward improvements in economic performance, but not efforts to sustain those improvements. So a few months after the improvements were realized, the plant's new level of economic performance was used as the new performance baseline, and the operations team's efforts were taken for granted. This is a major cultural issue. Achieving an improvement in performance warrants recognition, but sustaining improvement warrants equal recognition. The Dynegy Midstream Services management team recognized the importance of a change in culture that supported the effective utilization of DPMs. Their awareness represents one example of how a culture change can help.

Another issue that arose in several instances was an inclination by plant engineering teams to build dashboard displays that looked exactly like DPM dashboard displays. They then would populate them with a measure they could derive that had an associated economic value. This practice was very counterproductive in most cases. Not only did the measures identified not support the plant's manufacturing strategy; they encouraged behaviors that actually reduced the economic performance of the plant. Plant operators, and most employees, will typically strive to do their best for their organization. If they are provided with a set of measures that define what "good" is, they will try to perform to them. Driving performance according to the incorrect measures can lead to diminished economic performance.

The companies whose operations performed the best had devised a reward system that was integrated with the performance measures. Having a set of interesting displays is nice, but if companies want their employees to drive performance the best way to communicate this is with a reward system. Stronger rewards imply a higher emphasis placed on improvement and typically correlate with greater performance improvements. It is very important, however, for the management team to really consider what behaviors it wants and to structure the rewards to match those behaviors. For example, if only the improvements in performance are rewarded and not the sustaining of those improvements, unusual behaviors can result.

Finally, when the company had put performance measurement systems in place and made them operational and when it had identified the improvements, the plant operating teams often went to the firm's financial department to validate the economic value of the improvement. This proved to be a long and tedious task because today's cost accounting systems do not make it easy to isolate and identify economic improvements that occur within an area or process unit of a plant. Most plants at which the DPM projects were implemented used an enterprise resource planning (ERP) system as the basis for their financial reporting. The ERP systems did not provide the level of information necessary to confirm or deny the economic value of the improvements. This is a very key issue. If the financial department does not recognize that an economic improvement has occurred, in many companies the improvement is assumed not to exist. In their book *Cost and Effect*, Dr. Robert Kaplan and Dr. Robin Cooper address the shortcomings of today's cost accounting systems especially with respect to manufacturing performance. They propose a new model for cost accounting that combines ERP with activity-based management (ABM) systems as well as with operational and strategic performance measurement systems. Implementing the approach they recommend would tie the DPMs into a comprehensive accounting structure that will make operational decision support much more effective and align the firm's financial reporting with the actual measures of plant-level performance. This will lead to much more effective decision-making and economic management for manufacturing operations.

The advent of new approaches to performance measurement such as DPMs are steps along a new path leading to improved manufacturing performance. Many other issues need to be addressed if manufacturing operations are to gain the full positive impact of these approaches. Changes in corporate cultures and in cost-accounting approaches are supportive of a total performance-based organization. Implementing these changes will not be easy. Those manufacturers that take these new approaches very seriously and drive their manufacturing operations to new levels of economic performance will be the leaders in the new manufacturing economy.

NOTES

1. Fiske, Tom, "Improving Operational Performance by Integrating Technology, Information, and Culture," *ARC Insights,* March 14, 2001, page 1. *ARC Insights* is published by Automation Research Corporation of Dedham, MA.
2. Ibid, page 3.
3. Ibid, page 4.
4. Ibid, page 4.
5. Ibid, page 4.
6. Ibid, page 4.
7. Ibid, page 1.
8. Ibid, page 4.

Bibliography

A Special Report on Factory Automation, "Automation's Global Game," *IndustryWeek,* June 20, 1988.

A Special Report on Factory Automation, "Linking the Pieces of CIM," *IndustryWeek,* March 23, 1987.

A Special Report on Factory Automation, "The U.S. and Quality: A New Culture," *IndustryWeek,* April 17, 1989.

A State-of-the-Art Report, "Factories of the Future," *IndustryWeek,* March 21, 1988.

Abel, Janice, and Peter Martin, "Automation, Business, and Operating Advances Align into New Paradigm for Economic Performance Improvement," *Pharmaceutical Engineering,* January/February 2000.

Allen, L., and D. Stovicek, "CIM Goes to School," *Automation,* May 1990.

Anderson, Norman A., *Instrumentation for Process Measurement and Control,* Radnor, PA: Chilton, 1980.

Andrews, W G., and H.B. Williams, *Applied Instrumentation in the Process Industries,* Houston, TX: Gulf Publishing, 1979.

Babb, Michael, "Process Control Systems in the 1990s Will Enter a 'New Age'," *Control Engineering,* January 1990.

Badavas, Dr. Paul C., and Dr. Albert D. Epperly, *Statistical Process Control Integrated with Distributed Control Systems,* 1988 NPRA Computer Conference, Pittsburgh, PA, October 1988.

Badavas, Dr. Paul C., and Dr. Albert D. Epperly, *Statistical Process Control Integrated with Distributed Control in an Oven Industrial System,* white paper, The Foxboro Company, 1988.

Baer, Tony, "The Software Side of Quality," *Managing Automation,* May 1990.

Balch, Burton, and Ted Miller, "Statistical Process Control in Food Processing," *Intech,* July 1990.

Beaverstock, Malcolm, and Lorne Byzyna, "Measuring Performance Dynamically," *Control Engineering,* June 1990, volume 11.

Behar, M. E., "History in the Making," *Instruments,* March 1948.

Behar, M. E., "On Its Twentieth Anniversary Instruments Honors Instruments," *Instruments,* vol. 21, 1948.

Benassi, Frank, "The Long Road to Quality," *Managing Automation,* May 1990.

Berliner, Callie, and James Brimson, *Cost Management for Today's Advanced Manufacturing,* Boston, MA: Harvard Business School Press, 1988.

Bhote, Kevi R., "America's Quality Health Diagnosis: Strong Heart, Week Head." *Management Review,* May 1989.

Bowen, Earl K., *Statistics with Applications in Management and Economics,* Homewood, II.: Richard D. Irwin, 1960.

Box, George E. P., William G. Hunter, and J. Stuart Hunter, *Statistics for Experimenters: An Introduction to Design, Data Analysis, and Model Building,* New York: John Wiley & Sons, 1978.

Brimson, James A., "How Advanced Manufacturing Technologies Are Reshaping Cost Management," *Management Accounting,* March 1986.

Brown, Mark E., Kevin Parker, and Cory P. Senyard Jr., "Computer Integrated Manufacturing in the Chemical Process Industries," *Chemical Processing,* October 1989.

Bunnell, John, and Richard L. McAllister, "Continuous Improvement through Plantwide Integration," *Control Engineering,* June 1990, volume 2.

Bylinsky, Gene, "Challengers Move in on ERP," *Fortune,* November 22, 1999.

Cargill, Carl F., *Information Technology Standardization,* Maynard, MA: Digital Press, Digital Equipment Corporation, 1989.

Casti, John L., and Robert E. Larson, *Principles of Dynamic Programming - Part I-Basic Analytical and Computational Methods,* New York: Marcel Dekker, 1978.

Clark, Kim B., and Robert H. Hayes, "Why Some Factories Are More Productive Than Others," *Harvard Business Review,* September-October 1986.

Cooper, Robin, and Robert S. Kaplan, "Measure Costs Right: Make the Right Decisions," *Harvard Business Review,* September-October 1988.

Copi, Irving M., *Symbolic Logic,* Toronto: MacMillan, 1970.

Cox, Brad J., *Object Oriented Programming, An Evolutionary Approach,* Reading, MA: Addison-Wesley, 1986.

Crosby, Philip B., *Quality Is Free: The Art of Making Quality Certain,* New York: McGraw-Hill, 1979.

DeMarco, Tom, *Structured Analysis and System Specification,* New York: Yourdon, 1978.

Deming, W. Edwards, *Out of the Crisis,* Cambridge, MA: MIT, Center for Advanced Engineering Study, 1986.

Dixon, J. R., A. J. Nanni, and T. E. Vollmann, *The New Performance Challenge: Measuring Operations for World Class Competition,* Business One-Irwin, 1990.

Dobyns, Lloyd, "Ed Deming Wants Big Changes, and He Wants Them Fast," *Smithsonian,* August, 1988.

Drucker, Peter F., "The Emerging Theory of Manufacturing," *Harvard Business Review,* May-June 1990.

Drucker, Peter F., *Innovation and Entrepreneurship: Practices and Principles,* New York: Harper & Row, 1985.

Drucker, Peter F., *The Practice of Management,* New York: Harper & Row, 1954.

Drucker, Peter F., Robert G. Eccles, Joseph A. Ness, Thomas G. Cucuzza, Robert Simons, Antonio Davilla, Christopher Meyer, Robert S. Kaplan, and Davis P. Norton, *Harvard Business Review on Measuring Corporate Performance,* Boston: Harvard Business School Publishing, 1998.

Elwart, Steven P., and Peter G. Martin, "New Software Structures Extend Control Capabilities," *Control Engineering,* volume 2, June 1990.

Epperly, A. D., and P. C. Badavas, "Statistical Quality Control, Statistical Process Control: Part 1, Quality Improvement; Part 2, Participative Management; Part 3, Statistical Tools; Part 4, Problem Solving," *State-of-the-Art Newsletter,* Sec. 3 Nos. 5-8, Foxboro, MA: The Foxboro Company, 1987.

Ferguson, Marylyn, *The Aquarian Conspiracy,* Los Angeles: J. P. Tarcher, 1980.

Fiske, Tom, "Improving Operational Performance by Integrating Technology, Information, and Culture," *ARC Insights*, Automation Research Corp., Dedham, MA, March 14, 2001.

Franson, Deborah L., "A Roundtable Discussion on Automation Systems on the 1990's," *Control Engineering*, volume 2, June 1990.

Gagne, James, "Quality Performance Means More at Dow," Midland, MI: Dow Chemical U.S.A.

Ghosh, Asish, and Howard P. Rosenof, *Batch Process Automation: Theory and Practice,* New York: Van Nostrand Reinhold, 1987.

Goldstein, Mark; William Pat Patterson, and John Teresko, "Linking the Pieces of CIM," *IndustryWeek,* March 23, 1987.

Gordon, Maureen, E., "Manufacturing Automation Protocol (MAP)," *State-of-the Art Newsletter,* Foxboro, MA: The Foxboro Company, January 1987.

Gould, Lawrence, "The CIM Task Force: Purchasing CIM Is Also a Distributed, Hierarchical, Pursuit," *Managing Automation,* December 1990.

Grant, Eugene L., and Richard S. Leavenworth, *Statistical Quality Control,* 5th ed., New York: McGraw-Hill, 1980.

Harbor Research, "Monthly Outlook, CIM Failures," *Outlook*, Harbor Research Corp., July 1987.

Harbor Research, "Monthly Outlook, CIM Failures, II," *Outlook,* Boston, MA: Harbor Research Corp., August 1987.

Harbor Research, "Performance Measurement: Impact on Competitive Performance," Outlook, Boston, MA: Harbor Research Corp., volume 6, number 4, 1990.

Harrold, Dave, "Enterprise Integration Requires Understanding of the Plant Floor," *Control Engineering*, February 2000.

Harry, Mikel J., Ph.D., *The Vision of Six Sigma: A Roadmap for Breakthrough*, Tempe, Arizona: Six Sigma Academy, 1994.

Hays, William L., *Statistics for the Social Sciences,* New York: Holt, Rinehart and Winston, 1973.

Henderson, Bruce A., and Jorge L. Larco, *Lean Transformation,* Richmond, Virginia: The Oaklea Press, 1999.

Hickey, J., "Demands & Requirements Companies Will Face in the 1990's," *Presentation,* May 1990.

Hidden, A. E., and E. Lowe, *Computer Control in the Process Industries,* London: Peter Peregrinus Ltd., 1971.

Hillier, Frederick S., and Gerald J. Lieberman, *Operations Research,* San Francisco: Holden-Day, 1974.

Hiromoto, Toshiro, "Another Hidden Edge-Japanese Management Accounting," *Harvard Business Review,* July-August 1988.

Horn, George, "Japan: Should We Copy to Compete?" *Automation,* May 1990.

IBC Technical Services, *Computer Integrated Manufacture in the Process Industries-Conference,* London: IBC Technical Services Ltd., 1990.

Jasany, L.C., "Knowledge (and Power) to the People," *Automation,* July 1990.

Johnson, H. Thomas, and Robert S. Kaplan, "The Rise and Fall of Management Accounting," *Management Accounting,* January 1987.

Jones, Daniel T., Daniel Roos, and James P. Womack, *The Machine That Changed the World: The Story of Lean Production*, New York, Harper Perennial (Harper-Collins), 1991.

Juran, Dr. Joseph, *Juran on Leadership for Quality - An Executive Handbook,* New York: Free Press, 1989.

Juran, Dr. Joseph, *Juran on Quality Leadership,* videotape, Juran Institute, 1988.

Juran, J. M., *Managerial Breakthrough: A New Concept of the Manager's Job.* New York: McGraw-Hill, 1964.

Kaplan, Robert S., "Accounting Lag: The Obsolescence of Cost Accounting Systems," *Harvard Business School 75thAnniversary Colloquium on Productivity and Technology,* Cambridge, MA: Harvard Business School, March 1984.

Kaplan, Robert S., "One Cost System Isn't Enough," *Harvard Business Review,* January-February 1988.

Kaplan, Robert S., "Yesterday's Accounting Undermines Production," *Harvard Business Review,* July-August 1984.

Kaplan, Robert S., and Robin Cooper, *Cost & Effect, Using Integrated Cost Systems to Drive Profitability and Performance,* Boston: Harvard Business School Press, 1998.

Kaplan, Robert S., and David P. Norton, *Translating Strategy into Action: The Balanced Scorecard,* Boston: Harvard Business School Press, 1996.

King, Bob, *Better Designs in Half the Time: Implementing QFD-Quality Function Deployment in America,* Lawrence, MA: Goal/QPC, 1987.

Kompass, E. J., "The Road to Plantwide Information," *Control Engineering,* June 1990, volume 2.

Kuhns, William, Lewis Mumford, Siegfried Giedion, Jaques Ellul, Norbert Wiener, R. Buckminster Fuller; Marshall McLuhan, Harold Adams Innis, *The Post-Industrial Prophets Interpretations of Technology,* New York: Harper & Row, 1973.

Lareau, Albert F., "Bringing Information to Automation," *Control Engineering,* June 1990, volume 2.

MacGregor, John F., "On-Line Statistical Process Control," *Chemical Engineering Progress,* October 1988.

Mandl, Vladimir J., "Teaming Up for Performance," *Manufacturing System,* June 1990.

Martin, Peter G., "A New Blueprint for Automation," *Chemical Plants and Processing*, September 1989.

Martin, Peter G., "A Software Structure for Open Industrial Systems," *Proceedings of the Fourteenth Annual Advanced Control Conference,* Purdue University and Control Engineering, September 1989.

Martin, Peter G., "Automation Payback," *Industrial Computing,* Research Triangle Park, North Carolina: ISA—The Instrumentation, Systems, and Automation Society, August 1999.

Martin, Peter G., "Computer Control of Batch Processes," *Measurement and Control,* issue 106, September 1984.

Martin, Peter G., "Designing a Batch Control System," *I&CS,* October 1987.

Martin, Peter G., *Dynamic Performance Management: The Path to World Class Manufacturing,* New York: Van Nostrand Reinhold, 1993.

Martin, Peter G., "IEEE 802 Standards," *State-of-the-Art Newsletter,* Foxboro, MA: The Foxboro Company, July 1986.

Martin, Peter G., "Offene Industriesysteme fur die Verfarhrenstechnik-Schritt in ein neues Automatisierungszeitalter," *Verfahrenstechnik,* Nr. 10, 1989.

Martin, Peter G., "OIS: A Blueprint for Automation," *Control Engineering,* February 1989.

Martin, Peter G., "Open Communications in the Process Industries," paper presented at ISA—The Instrumentation, Systems, and Automation Society, Houston Channel Section, April 1990.

Martin, Peter G., "Open Industrial Systems," *Measurement and Control,* October 1989.

Martin, Peter G., "Reference Model of Open System Interconnection," *State-of-the Art Newsletter,* Foxboro, MA: The Foxboro Company, November 1985.

Martin, Peter G., "The Move to Open Communications in the Process Industries," paper presented at ISA—The Instrumentation, Systems, and Automation Society, Calgary Section, April 1990.

Martin, Peter G., "The Move toward Open Industrial Systems," *Proceedings of the Advanced Control Symposium,* Texas A&M University, February 1989.

Martin, Peter G., "The Move toward Standard Operating Systems," *State-of-the-Art Newsletter,* Foxboro, MA: The Foxboro Company, January 1986.

Maskell, Brian H., "Performance Measurement for World Class Manufacturing, Part I," *Manufacturing Systems,* July 1989.

Maskell, Brian H., "Performance Measurement for World Class Manufacturing, Part II," *Manufacturing Systems,* August 1989.

McClenahen, John S., "Automation's Global Game," *IndustryWeek,* June 20, 1988.

McLuhan, Marshall, *The Mechanical Bride: Folklore of Industrial Man.* Boston, MA: Beacon Press, 1952.

McLuhan, Marshall, and Quentin Fiore, *The Medium Is the Message: An Inventory of Effects,* New York: Bantam Books, 1967.

McSween, Terry, and Victor Zaloom, Jr., "Creating a Positive Work Environment," *Chemical Engineering,* June 1990.

Miller, Jeffery G., Alfred J. Nianni, and Thomas E. Vollmann, "What Shall We Account For?" *Management Accounting,* January 1988.

Miller, Jeffery G., and Thomas E. Vollmann, "The Hidden Factory," *Harvard Business Review,* September-October 1985.

Murrill, Paul W., *Fundamentals of Process Control Theory,* Research Triangle Park, NC: ISA—The Instrumentation, Systems, and Automation Society, 1981.

Naisbitt, John, *Megatrends: Ten New Directions Transforming Our Lives,* New York: Warner Books, 1982.

Passino, Ralph, "Computers Spur Profitability at General Chemical," *Chemical Processing,* November 1990.

Pastor, E., "Networks: The Backbone of Integration," *Automation,* May 1990.

Peat, F. David, *Artificial Intelligence: How Machines Think,* New York: Simon & Schuster, 1985.

Peters, Thomas J., and Robert H. Waterman Jr., *In Search of Excellence: Lessons from America's Best-Run Companies,* New York: Harper & Row, 1979.

Quak, Al, "Batch Control Systems," *State-of-the-Art Newsletter,* Foxboro, MA: The Foxboro Company, May 1986.

Qualtec, Inc., *Managing Quality Improvement, Participant Workbook,* Miami, FL: Florida Power & Light Co., 1988.

Qualtec, Inc., *Team Leader Training Course, Participant Workbook,* Miami, FL: Florida Power & Light Co., 1987.

Rheingold, Howard, *Tools for Thought.- The History and Future of Mind-Expanding Technology - The People and Ideas Behind the Next Computer Revolution,* New York: Simon & Schuster, 1985.

Rohan, Thomas M., "Factories of the Future," *IndustryWeek,* March 21, 1988.

Rossi, Dennis A., "Consider Integrated Solution for Total Quality," *Intech,* January 1990.

Schonberger, Richard J., "Customer Chains: Links to Survival," *IndustryWeek,* April 16, 1990.

Semler, Ricardo, "Managing without Managers," *Harvard Business Review,* September-October 1989.

Sentell, Gerald D., "Satisfying Customers: Quality, Culture, Paradigms and Management," Alcoa, TN: Tennessee Associates.

Shaw, William T., *Computer Control of Batch Processes,* Cockeysville, MD: EMC Controls, 1982.

Shewart, W. A., *Economic Control of Quality of Manufactured Product,* New York: Van Nostrand Company, 1931.

Shinskey, F G., *Process-Control Systems,* New York: McGraw-Hill, 1979.

Skinner, Wickham, "The Productivity Paradox," *Harvard Business Review,* July-August 1986.

Southard, Robert K., "Local Area Networks - An Overview, Part I," *Manufacturing Systems,* June 1990.

Special Edition, "Manufacturing Automation Protocol," *Control Engineering,* October 1985.

Special Factory Automation Edition, *I&CS,* March 1986.

Special 1991 Bonus Issue, "The Quality Imperative, What It Takes to Win in the Global Economy," *BusinessWeek,* October 25, 1991.

Spiegel, Murray R., *Probability and Statistics,* New York: McGraw-Hill, Schaum's Outline Series, 1975.

Spiegel, Murray R., *Statistics,* New York: McGraw-Hill, Schaum's Outline Series, 1961.

Stalk, George, Jr., "Time-The Next Source of Competitive Advantage," *Harvard Business Review,* July-August 1988.

Sweeny, Allen, *Accounting Fundamentals for Nonfinancial Executives and Managers*, New York: McGraw-Hill, 1972.

The Foxboro Company, *Introduction to Process Control,* Foxboro, MA: 1986.

The MAP Process Industries Initiative Working Group, *AW in the Process Industry,* Dearborn, Michigan, MAP/TOP Users Group of the Society of Manufacturing Engineers.

Walsh, Susan, "Perspectives on the Age of the Smart Machine," *Managing Automation,* December 1989.

Waterman, Robert H., *Adhocracy.- The Power to Change: How to Make Innovation a Way of Life.* Knoxville, TN: Whittle Direct Books, 1990.

Watts, Patti, Section Editor, "Special Section: In Search of Quality," *Management Review,* May 1989.

Williams, Theodore J., *A Reference Model for Computer Integrated Manufacturing (CIM): A Description from the Viewpoint of Industrial Automation,* Research Triangle Park, NC: ISA—The Instrumentation, Systems, and Automation Society, 1989.

Zuboff, Shoshana, *In the Age of the Smart Machine,* New York: Basic Books, 1988.

Index